The Cytology and Life-History of Bacteria

K. A. Bisset, D.Sc.
Head of the Department of Bacteriology,
University of Birmingham

THIRD EDITION

E. & S. LIVINGSTONE
EDINBURGH AND LONDON, 1970

SBN 443 00660 1

Printed in Great Britain

Preface to Third Edition

Since the first and second editions of this book were published, there has been a revolution in the attitude of microbiologists and others to the study of bacterial cytology. At one time the rarest of pursuits, it has become relatively commonplace. The nuclear cytology of *Escherichia coli* and the electron microscopy of ultra-sections of the cell envelopes and mesosomes of sporing bacilli might be called specialist subjects in their own right. Preparations showing cellular structure in Gram-positive bacilli and cocci are made and studied by undergraduates.

There has been an enormous increase in available information, but there has not truly been a revolution in the subject itself, since the greatest service of the new observations has been to confirm, as well as to extend and elaborate, what was already known, or believed to be true. For this reason, I thought it necessary to show the older and newer work in perspective, by adopting a more historical approach than in previous editions. And because the literature has grown so greatly in volume, I have employed the convention of numbered references in the text, instead of listing them at the head of each section.

K. A. BISSET

1970

Acknowledgements

I wish to acknowledge with gratitude the assistance given to me in the production of this book by the Editors of the *Journal of Hygiene*, the *Journal of Applied Bacteriology*, the *Giornale di Microbiologia*, the *Journal of General Microbiology*, the *Journal of Pathology and Bacteriology*, *Experimental Cell Research*, and *Biochimica et Biophysica Acta*, all of whom lent blocks of illustrations as well as giving permission to reproduce others; by the *Cambridge University Press*, *Oliver & Boyd Ltd.*, the *Academic Press* and the *Elsevier Press* who provided the blocks in question, and especially by Dr Emmy Klieneberger-Nobel, Dr Woutera van Iterson, the late Professor J. Tomcsik, Dr A. L. Houwink, Drs Birch-Anderson, Maaløe and Sjöstrand, Dr C. D. Beaton, Dr Germaine Cohen-Bazire, Mr A. A. Tuffery, Dr P. D. Walker and his collaborators, whose admirable micrographs are the subject of many of these illustrations.

I wish to also thank the Editors of the *Proceedings of the Royal Society*, the *Journal of Bacteriology*, the *Ergebnisse der Mikrobiologie*, the *Journal of Biophysical and Biochemical Cytology*, the *Annales de l'Institut Pasteur*, the *Proceedings of the Society for Experimental Biology and Medicine* and *Cold Spring Harbor Symposia*, Professor R. H. Stoughton, Professor R. J. V. Pulvertaft, Dr J. Lederberg, Dr Phyllis Pease, Dr H. A. Bladen, Dr D. E. Bradley, Dr C. F. Robinow, Dr L. J. Rode, Dr C. C. Remsen, Dr Roger Cole, Dr D. G. Lundgren, Drs J. W. Czekalowski and G. Eaves, Drs E. L. Wollman, F. Jacob and W. Hayes, Dr M. H. Jeynes, Dr Joyce Grace, Dr R. R. Mellon, Professors E. O. Morris and A. R. Prévot, Drs Chapman and Hillier, and Mr L. O. White for permission to reproduce illustrations.

Lastly, I wish to thank Mrs Caroline Hale-McCaughey, F.I.M.L.T., for her skilled and patient work in the preparation of the cytological material upon which all three editions of this monograph have been based.

K. A. Bisset

1970

Preface to First Edition

This book does not attempt to review the literature upon bacterial cytology, of which the bulk is very great and the value, in many cases, difficult to assess. The bibliography is confined to a relatively small number of works, almost all recent. No attempt has been made to supply references for analytical discussion or general information.

The purpose is rather to present a reasoned case for regarding bacteria as living cells with the same structure and functions as other living cells, and to correlate the available information upon the various types of bacteria.

Bacteria, as living creatures, have been little studied. It is their activities as biochemical or pathological agents which have received almost undivided attention. Even these problems, however, cannot fail to be clarified by a better knowledge of the organisms responsible.

It is also hoped that biological workers in other fields may profit by contact with this, largely unknown, body of evidence, and may find the comparisons and analogies useful and stimulating in their related studies.

I have attempted, as far as possible, to base my arguments upon my own observations, or upon such information as I have been able personally to confirm. Where I have not had the opportunity to do so, I have tried to indicate clearly the status of the argument.

K. A. B.

December 1949

Contents

Illustrations

CHAPTER 1

Introduction

Despite the very great advances in our knowledge of the cytology of bacteria that have taken place in the last twenty-five years, it remains true that much of what is written about them is fallacious, and is based upon the assumption that, because of their small size, and the difficulty, by the methods usually employed, of observing the complexities of their structure, they may be regarded as simple in form and primitive in phylogeny.

The temptation to regard small size and simplicity of structure, whether real or apparent, as criteria of a primitive condition, has often proved the cause of error and confusion in the classification of other groups of living organisms. As more information becomes available it is almost invariably discovered that the simplest creatures exhibit characters which suggest a relationship with others, much more complex, or may themselves prove to be less simple than had been believed. This has proved to be true of bacteria also. Although for long claimed, in spite of much evidence to the contrary, to be almost structureless cells, reproducing by simple fission, they have proved to possess an intricacy of structure rivalling that of any other type of living cell, and to undergo life-cycles of considerable complexity.

For a long time, much more was learned of the physiology of bacteria than of their morphology, and this was mainly because of their importance in medicine, industry and agriculture. The immediate, practical problems of bacteriology have overshadowed the more academic questions of their biological nature. The techniques which were devised for the solution of these problems were notable, in almost every case, for their failure to provide even a minimum of basic, biological information. Indeed it may be said that much of the information of this nature, accumulated during the first forty years of systematic bacteriology, tended rather to obscure than to clarify the underlying truths.

Especially is this true of the staining techniques employed for routine examination of bacteriological material and cultures. The distorted vestiges of bacteria that survived the technique of drying and heat-fixation were accepted as truly indicative of the morphology of the living organisms. And while, from

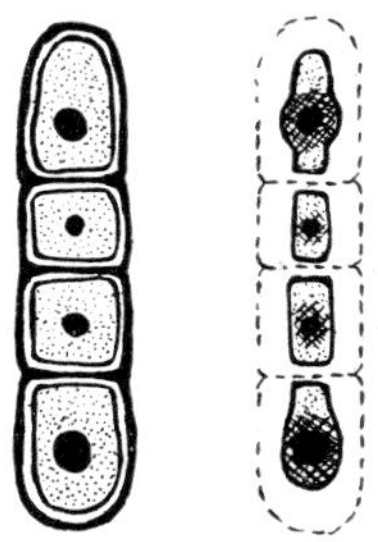

A B

(Reproduced from the Journal of General Microbiology)

FIG. 1

THE MORPHOLOGY OF *C. DIPHTHERIAE*

A. True morphology.

B. 'Typical appearance' in heat-fixed material. The cell contents are shrunken and the cell wall unstained.

time to time, satisfaction was expressed at the fact that bacteria will survive, undistorted, treatment which produces the most obvious damage in larger cells, the validity of the assumption that they did, in fact, survive such treatment was seldom called to question (Figs 1, 2 & 3).

Staining methods were devised, almost without exception, for the purpose of identifying clinically important species of bacteria, and were often most admirably suited to this task. It is surprising to find, however, that much time and labour has been directed to the elucidation of the appearances observed by these methods, and the explanation, in cytological terms, of the artefacts they produce. A classical example of this is the substance known as volutin, which is still the subject of some discussion and even research. [1] Originally given as a name to the large granules visible in *Spirillum volutans*, it has been applied to many other bacterial granules since 1904, when Meyer [2] illustrated two structures in a sporing bacillus. One of these was manifestly an immature endospore, the other, one of those granules at the junction of cell-wall and developing septum that has been called a growing point [3] or, more recently, a mesosome. [4, 5] The well-known appearance in *Corynebacterium diphtheriae* is illustrated in Figure 1. Here, the granules are simply the dried contents of entire, small cells in a strongly septate bacterium. [6]

The artificiality of recent views upon bacterial morphology thus served to widen the gap between bacteriology and other biological sciences, as well as to confuse and retard the advance of bacteriology itself.

In the evolution of modern cytological methods, much was owed to the interest taken by Polish mycologists in the myxobacteria. [7, 8, 9] These micro-organisms do not respond well to the techniques of heat-fixation and Gram's stain, most usually employed in routine bacteriology, and the necessity for the employment of more refined methods of examination has encouraged the study of eubacteria in a similar manner. The readily-demonstrable nuclear structures and beautiful and complex life-cycle of myxobacteria stimulated the search for the truth concerning the parallel structures and processes in those bacterial genera more commonly encountered in the laboratory, and especially in the parasitic species and important contaminants.

The study of bacterial cytology, like most other branches of science, has passed through periods of active development and of stagnation or even retrogression. After Schaudinn's [10, 11] accurate studies on cell division, made as early as the turn of the century, little attention was paid to the morphology of bacteria, until the thirties, when there was a sudden upsurge of interest in the subject. Using classical methods of staining and light-microscopy, a considerable proportion of our present knowledge of bacterial structure, down to the level of size that can be discerned by these means, was obtained about that time. The general description of vegetative cells given by Robinow in 1945 [12] has, in most respects, never been improved upon, and contains only one major error (the so-called peripheral nucleus of the endospore, which is an artefact). At this time, the electron microscope came into general use, but its contribution was confined, initially, to the study of the flagella and of some disrupted elements derived from the cell wall. [13, 14] So far as the flagella were concerned, it was exceedingly valuable because, in bacteria, the size of these structures is below the limits of resolution by visible light. The usefulness of the electron microscope in the elucidation of the internal structures of bacteria is severely limited by the opacity of the complete cells to the electron beam. Occasional attempts were made however, to correlate such electron-dense granules as could sometimes be observed, with the granules in similar positions that can be demonstrated by histochemical staining methods under the light-microscope.[1, 15] Information has also been obtained by the electron microscopy of carbon replica casts of bacterial surfaces[16] or by negative staining.[17]

Apart from these studies, and especially those concerning the bacterial flagella, and their mode of attachment to the cell wall,[18] the real contribution of the electron microscope to the study of bacterial cytology commenced with the development of sectioning methods, which permitted the internal structures to be examined at high magnifications (Fig. 4). The first studies of this type gave more information about the cell envelopes than about the more inward portions of the cell,[19, 20] but experience in methods, especially of fixation and embedding, soon wrought a considerable improvement, and brought about an advance in knowledge comparable with the achievements of classical microscopy in the previous two decades. This can be judged by comparing the 1956 and 1965 Symposia of the Society for General Microbiology.[21, 22] At the same time, it is worth remembering how much basic information was already on record, before this second revolution commenced. For example, Bergersen[23] in 1953 had already described that association between nucleus and mesosome which was rediscovered in 1964, by electron microscopy of ultra-sections,[24] and is now regarded as one of the greater advances that have been made in the co-ordination of information obtainable from cytological and genetical studies of the bacterial nuclear apparatus.

In fact, bacteria are now widely used for genetical studies, but it is unfortunate that an enormously high proportion of this work has been performed upon a single species, *Escherichia coli*. Nevertheless, the beautiful demonstrations of sexual conjugation in the vegetative phase[25, 26] and of the chromosomal thread[27] are owed to geneticists. It can only be a matter of time until more complete studies are made of other bacterial genera and of different stages in their life-cycles. It is especially disappointing that one of the best-recognised phenomena in the nuclear cycle of bacteria, the apparent process of autogamy and reduction in the maturation of the resting stage (spore or microcyst)[28, 29, 30, 31] has never attracted the attention of geneticists (Chaps 4 & 5).

Perhaps the greatest change that has taken place in bacterial cytology in recent years is that, from being the esoteric preserve of a few enthusiasts it has become a study of wide and general interest, and biochemists, who previously tended to ignore morphology and structure, now propose actual physical shapes for such entities as ribosomes, that are truly chemical concepts, and outside the scope of this book.[32]

REFERENCES

1. Duguid, J., Smith, I. W. & Wilkinson, J. F. (1954). *J. Path. Bact.* **67**, 289.
2. Meyer, A. (1904). *Bot. Ztg.* **62**, Abt. I, 113.
3. Bisset, K. A. (1951). *J. gen. Microbiol.* **5**, 155.
4. Fitz-James, P. C. (1965). *Symp. Soc. gen. Microbiol.* **15**, 369.
5. Walker, P. D. & Baillie, A. (1968). *J. appl. Bact.* **31**, 108.
6. Bisset, K. A. (1949). *J. gen. Microbiol.* **3**, 93.
7. Badian, J. (1930). *Acta Soc. Bot. Pol.* **7**, 55.
8. Badian, J. (1933). *Acta. Soc. Bot. Pol.* **10**, 361.
9. Krzemieniewska, H. (1930). *Acta Soc. Bot. Pol.* **7**, 507.
10. Schaudinn, F. (1902). *Arch. Protistenk.* **1**, 306.
11. Schaudinn, F. (1903). *Arch. Protistenk.* **2**, 421.
12. Robinow, C. F. (1945). Addendum to *The Bacterial Cell*, ed. Dubos, R. J. Harvard University Press.
13. Houwink, A. L. (1953). *Biochim. biophys. Acta*, **10**, 360.
14. Salton, M. R. J. & Williams, R. C. (1954). *Biochim. biophys. Acta*, **14**, 455.
15. Glauert, A. M. & Brieger, E. M. (1955). *J. gen. Microbiol.* **13**, 310.
16. Bradley, D. E. & Williams, D. J. (1957). *J. gen. Microbiol.* **17**, 75.
17. Bladen, H. A. & Mergenhagen, S. E. (1964). *J. Bact.* **88**, 1482.
18. van Iterson, W. (1947). *Biochim. biophys. Acta*, **1**, 527.
19. Chapman, G. B. & Hillier, J. (1953). *J. Bact.* **66**, 362.
20. Robinow, C. F. (1953). *J. Bact.* **66**, 300.
21. Symposium (1956). 6*th Symp. Soc. gen. Microbiol.*
22. Symposium (1965). 15*th Symp. Soc. gen. Microbiol.*
23. Bergersen. F. J. (1953). *J. gen. Microbiol.* **9**, 26.
24. Ryter, A. & Jacob, F. (1964). *Annls Inst. Pasteur, Paris*, **107**, 384.
25. Lederberg, J. (1956). *J. Bact.* **71**, 497.
26. Anderson, T. F., Wollman, E. L. & Jacob, F. (1957). *Annls Inst. Pasteur, Paris*, **93**, 450.
27. Cairns, J. (1963). *J. molec. Biol.* **6**, 208.
28. Badian, J. (1933). *Arch. Mikrobiol.* **4**, 409.
29. Klieneberger-Nobel, E. (1945). *J. Hyg., Camb.* **44**, 99.
30. Bisset, K. A., Grace, J. B. & Morris, E. O. (1951). *Expl Cell Res.* **2**, 388.
31. Ellar, D. J. & Lundgren, D. G. (1966). *J. Bact.* **92**, 1748.
32. Möller, W. (1969). *Nature, Lond.* **222**, 979.

CHAPTER 2

Technique

Most advances in the study of bacterial cytology are now made by the electron microscopy of ultra-sections, but this elaborate and expensive technique is not at everybody's disposal, and it is necessary for microbiologists to have some knowledge of simpler methods. It is also true that a much clearer idea of general morphology, rather than ultrastructure, can usually be obtained by light microscopy. In the early days of the subject, progress was considerably retarded by the fact, which may at first have appeared advantageous, that recognisable microscopic preparations of bacteria can be made by the technique of the heat-fixed film. A small quantity of a bacterial culture, or of pus, or similar pathological material, is thinly spread upon a slide, dried, and then heated strongly with a naked flame, in order to fix it firmly upon the slide. Bacteria fixed in this manner and stained by Gram's method, or simply with a strong solution of a basic dye, dried once more and examined directly under the oil-immersion lens of the microscope (the oil serving also as a clearing agent), preserve an appearance which enables them to be recognised as bacteria, and even classified within rather broad limits. Their appearance under this treatment has become familiar to generations of bacteriologists, and is usually that which is recorded in the descriptions of species. Little or no detail can be perceived in such a preparation, and equally little structure can, as a rule, be made out in unstained living bacteria, especially as these are seldom at rest, either because of their own motility or from the effect of Brownian movement. Even phase-contrast or electron microscopy of bacteria usually provide very little information unless the preparations are quite elaborately treated in advance.

The main reason for the uniform appearance of stained bacteria is that their affinity for the basic dyes which are commonly employed is so great that the strongly stained cytoplasm and cell membranes mask the underlying structures. The masking effect is accentuated by the shrinkage that results from drying. This shrinkage is often very considerable, reducing the bacterium to as little as half or a third of its natural size, and manifesting itself typically in the appearance of the anthrax bacillus or of related chain-forming bacilli, in which considerable gaps are seen between the visible bacilli, actually the shrunken protoplasts. The rigid cell wall remains unstained and invisible, holding the chain together.

Many bacteria are multicellular, and their appearance can be considerably altered by drying. The granular appearance of corynebacteria and mycobacteria is enhanced by the shrinkage of the short, spherical or even discoidal cells that make up the bacillus, so that unstained gaps appear between them. In this case also the cell wall remains unstained, but retains the dried cells in their original relationship (Figs 1 & 2).

It will thus be seen that there are three main problems to be solved in the demon-

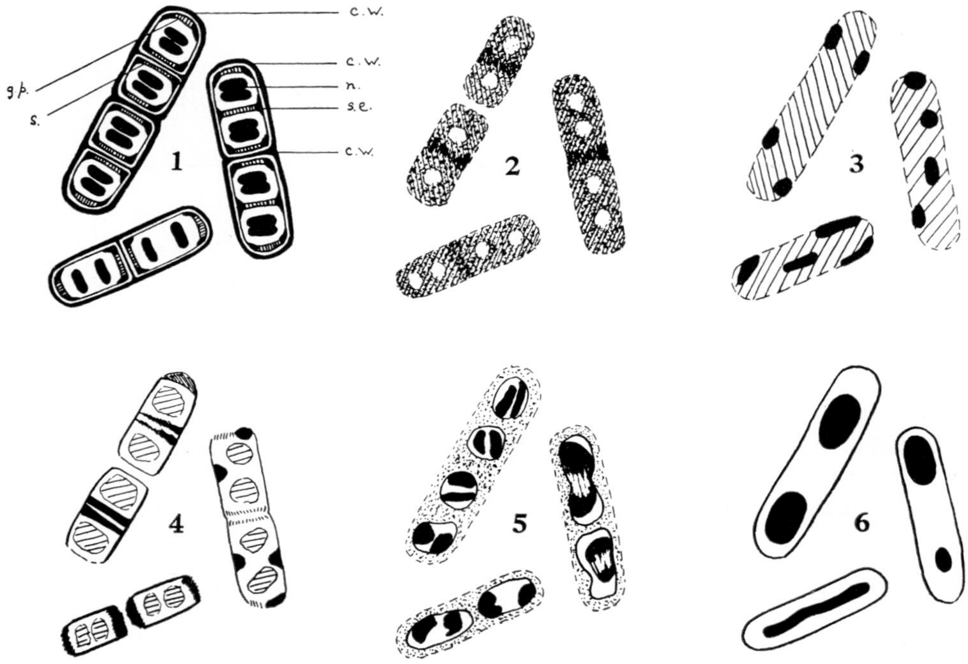

(*Reproduced from Experimental Cell Research*)

FIG. 2

A GROUP OF *BACILLUS* FIXED AND STAINED BY VARIOUS METHODS

(1) Diagram of the cytological structure of a group of bacilli at various stages of cell division. The top-left bacillus is divided into four cells by complete cross-walls; fission is commencing. The top-right bacillus is divided by a complete cross-wall and subdivided by cytoplasmic septa alone. The lower bacillus has recently divided and has only two cells. *c.w.*, cell wall; *g.p.*, growing point; *s.*, cytoplasmic septum preceding cross-wall; *s.e.*, septum in early stage; the darkly-stained junctions of the septa embody mesosomes; *n.*, nucleus.

(2) The same group demonstrated by a simple, basic dye.

(3) Stained by Gram and over-decolorised.

(4, 5) Artefacts caused by unsuitable fixation.

(6) Sporulation appearances.

stration of the true morphology of bacteria. Distortion due to drying must be avoided, the masking effect of the strongly staining protoplasm and cell membranes overcome, and those structures demonstrated which, like the cell wall, are difficult to stain. The first is simple and entails merely the avoidance of drying at all stages of preparation. The second and third present more difficulty. The problem of overcoming the masking effect of the surface structures was solved, as so often happens, by accident.

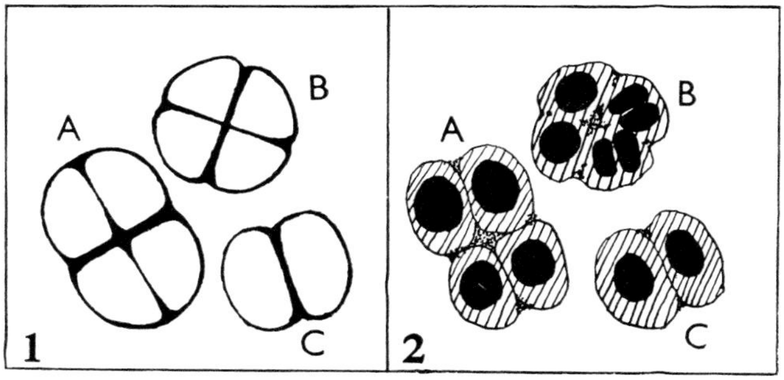

(*Reproduced from the Journal of Bacteriology*)

FIG. 3
THE CYTOLOGICAL STAINING OF COCCI

(1) *Micrococcus cryophilus* by Hale's method for cell walls. Showing two- or four-celled cocci.

(2) A similar group stained by trichloracetic acid and Giemsa. Such arrangements of nuclear material have been misconstrued as mitotic figures by some observers.

THE STAINING OF THE NUCLEUS

The Feulgen reaction is a microchemical test that depends upon the formation of a purple compound when aldehydes react with Schiff's reagent. A positive Feulgen reaction is given by deoxyribose nucleic acid, after its purine bases have been removed by acid hydrolysis. Ribonucleic acid does not give a positive reaction. The hydrolysis is performed in Normal hydrochloric acid at a temperature of 60°C., and the subsequent staining by Schiff's reagent reveals the nuclear structures of bacteria with reasonable clarity. This was one of the first methods to give a true picture of the bacterial nucleus, [1, 2, 3] and it was later discovered that if the final staining was performed with Giemsa's solution, instead of Schiff's reagent, a much clearer preparation was obtained. [2, 4, 5] This was the acid-Giemsa stain (Fig. 5), and it has been the basis of nearly all recent work upon the bacterial nucleus, ever since Robinow's beautiful photomicrographs first convinced bacteriologists of the reality of these structures, in 1942, although an accurate illustration of the nucleus, stained by Giemsa, had been published in 1919, [6] and acid-hydrolysis was first used for the purpose in 1924. [7]

The action of the preliminary treatment with hydrochloric acid is twofold; the nucleic acids of the nuclear structures are partially hydrolysed, the aldehyde group of the associated pentose sugar is released and combines with the staining agent; at the same time, the stainable material of the outer layers of the cell is more completely hydrolysed, so that its masking effect is reduced. [8,9]

To perform the stain, smear preparations are made upon slides or coverslips. They may be unfixed, although these tend to wash off, or fixed in osmic acid vapour. Most fixatives should be avoided as they can completely alter the appearance of the nucleus.

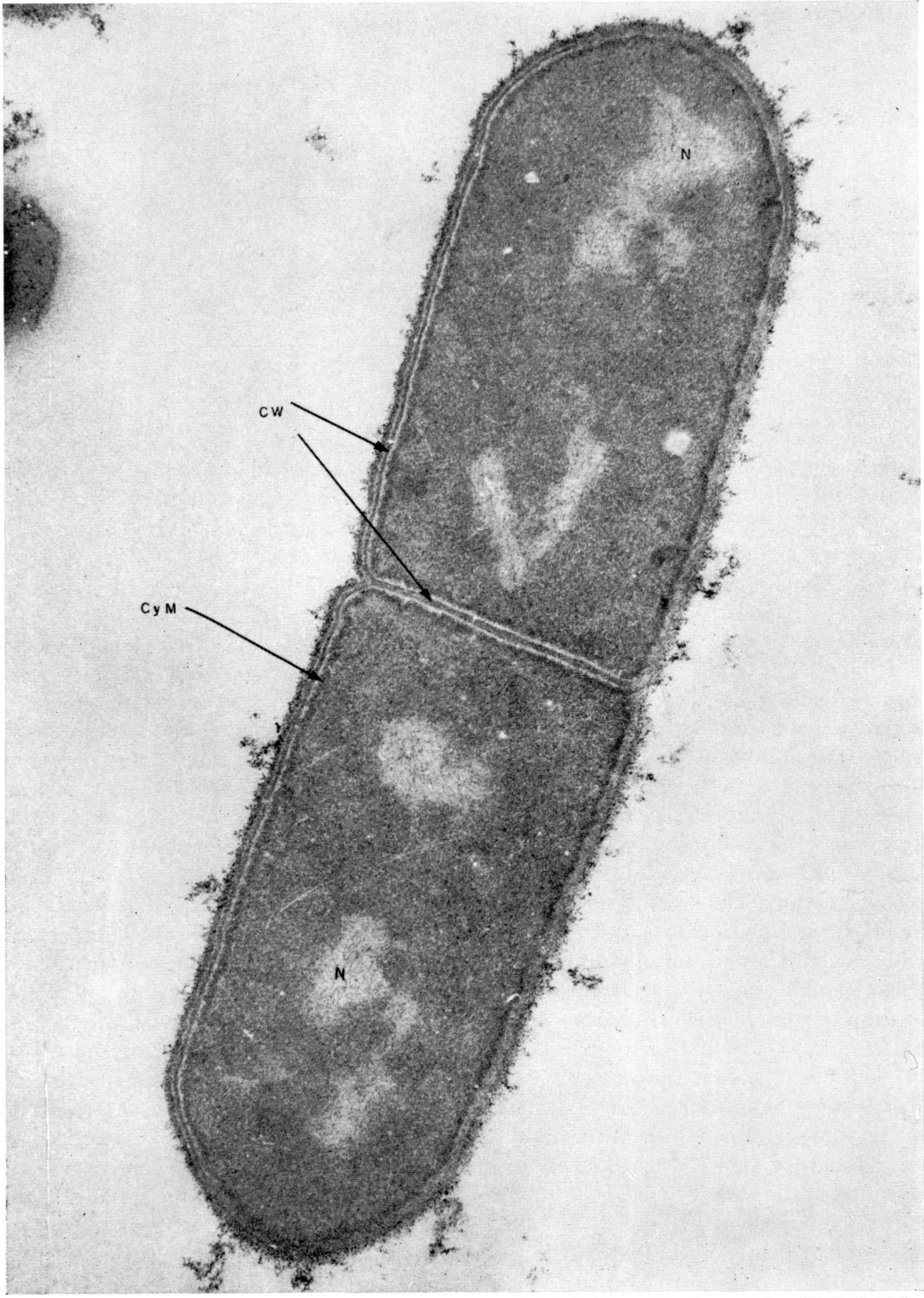

(*Reproduced from the Journal of Applied Bacteriology, by permission of Dr P. D. Walker*)

it is now recognised that the mesosomes (Chap. 3) show this activity. However, reagents such as triphenyltetrazolium or supposedly specific mitochondrial stains give very irregular results, and may show reactions in almost any site. [25]

The enigmatic nature of the majority of demonstrable granules in bacteria is tacitly admitted by the practice of coining for them such titles as 'metachromatic granules' or 'volutin'. These granules are rarely apparent except in dried material, and are often artefacts; although reserve foodstuffs, in the form of polysaccharides or lipids, may normally be present in the bacterial cell. It has been shown that their production is greatly influenced by nutritional conditions. [26, 27]

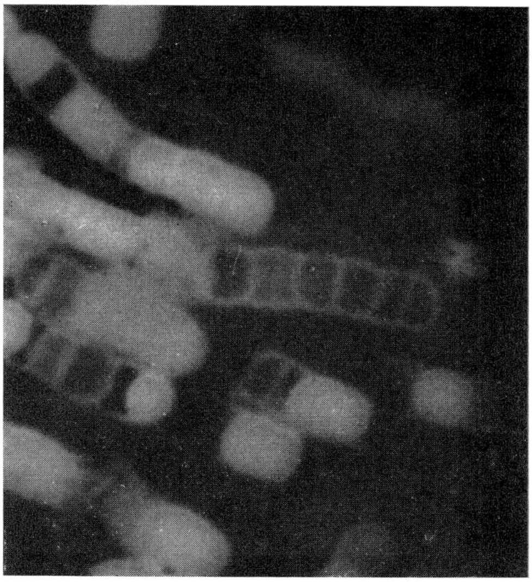

(Reproduced from Experimental Cell Research, by permission of Dr Claes Weibull and the Academic Press)

FIG. 7
MULTICELLULARITY

Bacillus megaterium. A simple preparation, made by crushing the bacilli gently between slide and coverslip and staining negatively with dilute indian ink. The very complex septate structure is clearly shown. ×3800.

CELL WALL STAINS

The bacterial cell wall is almost invariably quite invisible in ordinary microscopic preparations, although it is not at all difficult to demonstrate. It can be stained very easily with Alcian blue, [28] and in crushed cells the wall and cross-walls can be rendered visible by simple positive or negative staining [28, 29] (Fig. 7).

The most generally employed techniques make use of tannic acid (Gutstein[7]) or phosphomolybdic acid (Hale[31]) as preliminary mordants. These agents serve the dual purpose of mordanting the cell wall and of altering the protein material of the cell so that it is rendered unstainable and does not obscure the details of the transverse septa, where these occur (Fig. 8). Tannic acid forms a stainable complex upon the surface of the cell wall, and if it is stained before the mordanting process is complete, this complex may produce an outline picture but fail to show internal details. After treatment with 5-10 per cent tannic acid, the wall can be stained with 0·2 per cent crystal violet; using 1 per cent phosphomolybdic acid as a mordant, 1 per cent methyl green gives the clearest results, but crystal violet and a variety of other stains can be employed. The times required for staining and mordanting vary from a few seconds upwards. It has been suggested that these different methods do not all act upon the same part of the cell envelopes [32] and this may well be true.

The underlying cell membrane is not easily demonstrated. Transverse septa derived from it, containing a large protein component, are stainable by simple, basic dyes or by acid-

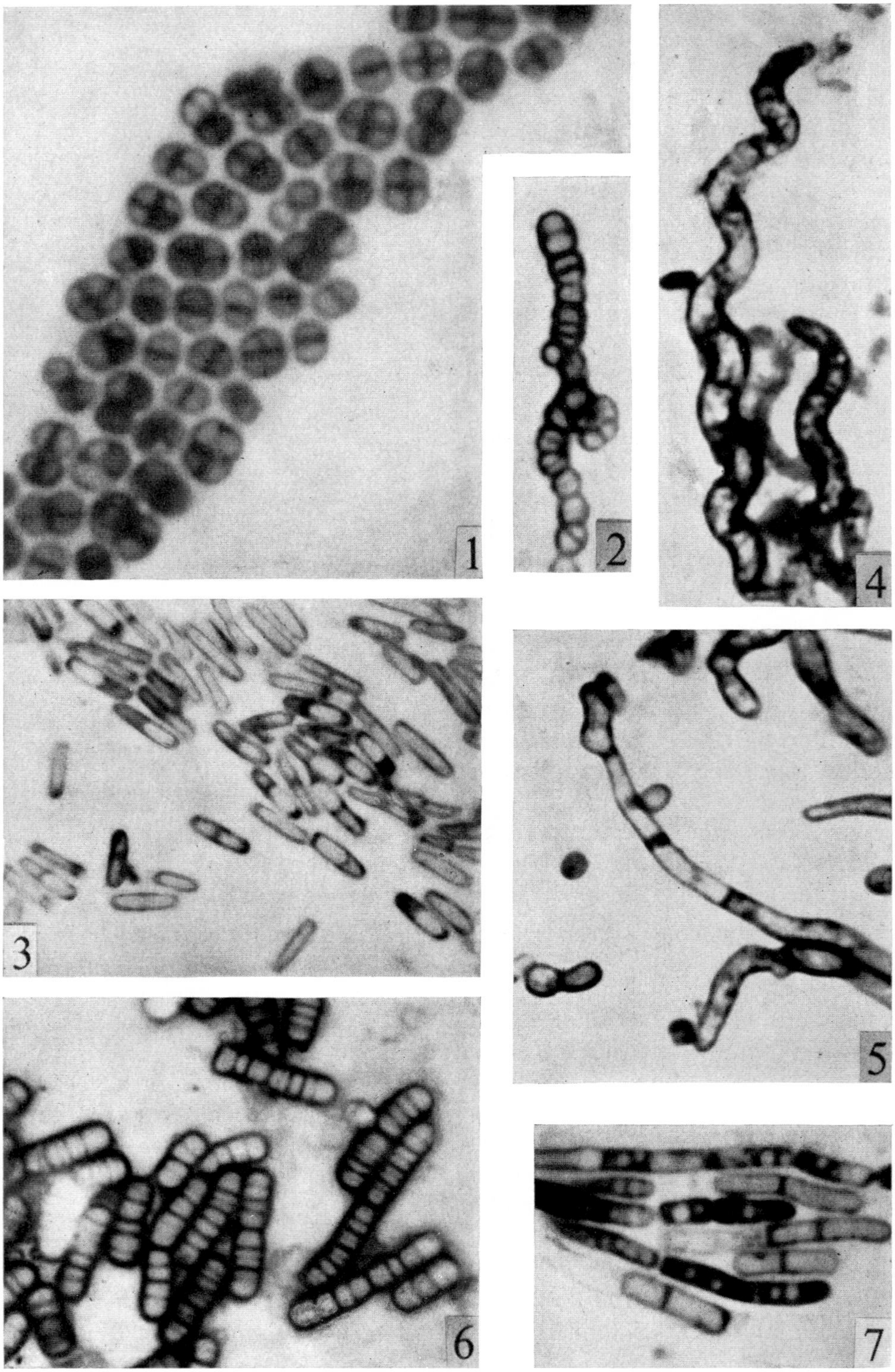

(*Plates 5, 6 & 7 reproduced from Experimental Cell Research*)

Giemsa, and are sometimes rendered more obvious by fixation with Bouin's solution or similar agents. Bacteria that have been slightly plasmolysed by such fixatives can be stained by 0·05 per cent Victoria blue in such a manner as to demonstrate the cell wall and cell membrane simultaneously [33] (Fig. 9).

Bacterial cell membranes are also stained by dyes of the Sudan, fat soluble group, [34] because of their lipid content.

The cell wall proper stains well by methods intended to demonstrate polysaccharides; [35] and most methods of staining the wall stain polysaccharide capsules also, to some extent. However, these and the classical capsular stains give no hint of the complex structure that can be seen by phase-contrast microscopy (Fig. 10), and they are of diagnostic rather than cytological interest.

THE MOUNTING OF MATERIAL

Cytological preparations should be mounted upon the thinnest available coverslips. It is a distinct advantage to prepare the smear upon the coverslip, so that the part of the preparation which is firmly adherent to the glass is nearest to the objective of the microscope and is not disturbed by Brownian movement. It is also simpler to transfer a coverslip from one reagent to another without the necessity of employing large volumes of fluid. Small volumes of reagents should be renewed at frequent intervals.

Coverslips may be sealed to the slide, at the edges, with beeswax or a mixture of paraffin wax and petroleum jelly, unless the preparations are dehydrated and mounted, to which there is no theoretical objection. In practice,

---

Fig. 8

CELL DIVISION IN BACTERIA

Cell division, in various bacterial types; stained by Hale's method. All at ×3000 except (1), ×4500, and (6) ×1700.

(1) Typical multicellular coccus. Each unit, which appears by routine staining methods as a single, spherical cell, contains two, four or more cells, separated by cross-walls, each of which is formed at right-angles to the preceding.

(2) *Streptococcus* sp. Although each coccus may contain two or more cells, the cross-walls are all in the same plane.

(3) Typical unicellular bacterium (*Pseudomonas* sp.). Dividing cells are separated by a short-lived septum, lacking the rigidity of a true cross-wall. Frequently one pole, the growing point, is marked by a concentration of stainable material.

(4) *Spirillum* sp. (modified stain by Mr R. A. Fox). The cross-walls of spirilla are not easy to demonstrate, nor, in view of their differences of plane, to photograph, but they appear to be true cross-walls, comparable with those of the Gram-positive genera.

(5) *Nocardia rhodnii*. Irregular septation with transient branching of the filaments.

(6) *Caryophanon latum*. The highest degree of multicellularity is seen in these giant, intestinal bacteria, where each cell is reduced to a disc

(7) An exceptional degree of irregular multicellularity is seen in aberrant stains of *Bacillus cereus*, and this appearance is accentuated by their high lipid content.

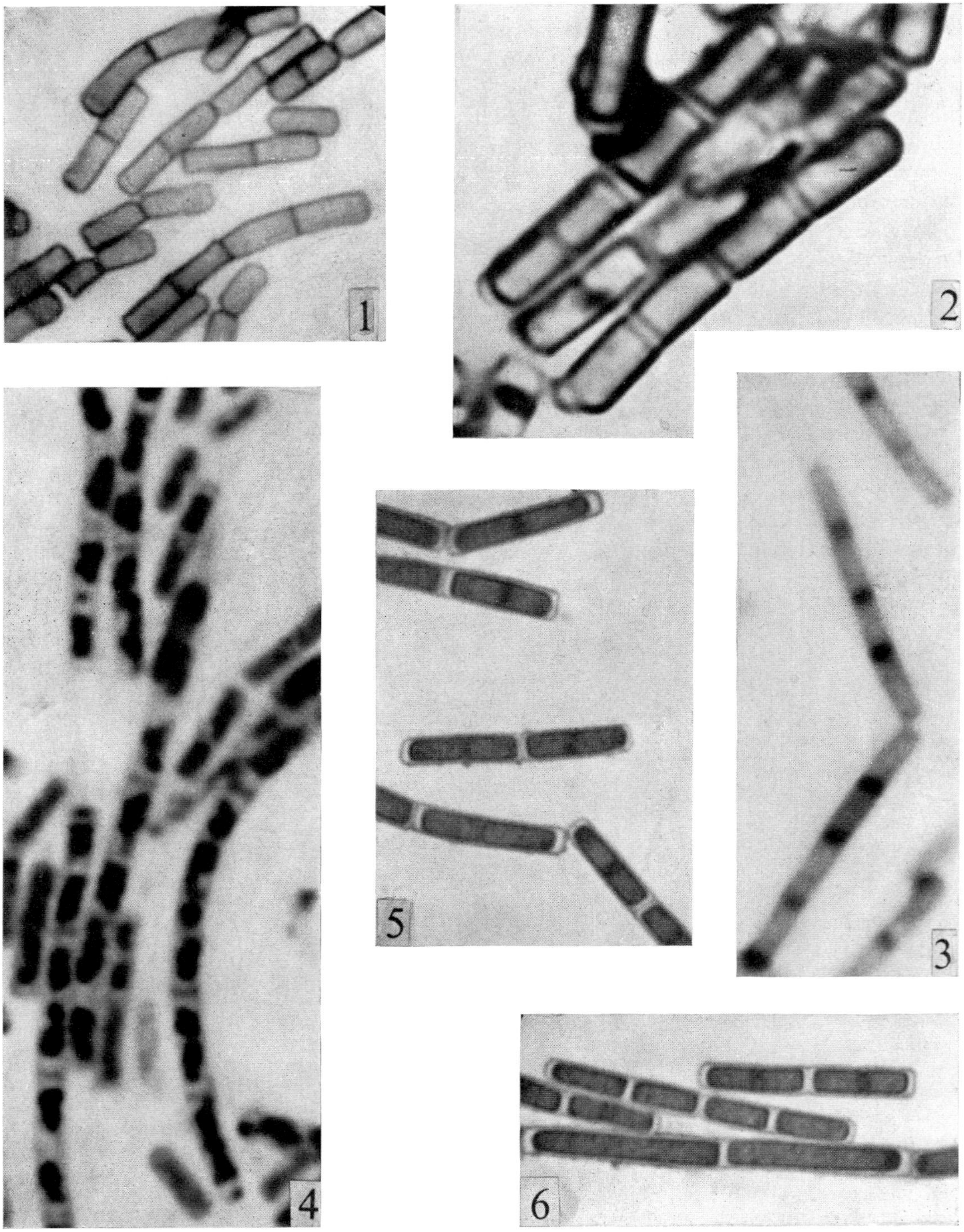

Plates 2, 3 & 4 *reproduced from the Journal of General Microbiology*)

however, it will be found that some shrinkage and distortion will usually result, and the clarity of the finished preparation will compare unfavourably with that of a simple water mount, although possessing the advantage of being permanent. If the edges are carefully sealed, a water mounted preparation will last for several days in the refrigerator, although it may deteriorate rapidly at room temperature.

It is worth emphasising that far more detail can usually be made out in a good photomicrograph, with all the advantages of colour filtration, than can be discerned, by the most experienced observer, in direct microscopic examination. Impermanence of preparations is thus of little importance provided that interesting appearances are photographed. It is also true that appearances which cannot be reproduced, more or less at will, are unlikely to be either true or important, and their impermanence is not to be regretted.

FLAGELLAR STAINS

The classical methods for staining flagella may be obtained from any elementary textbook upon practical bacteriology. A recently-devised method can be used to demonstrate flagella and cell wall structure simultaneously, [36] but is too difficult for routine use.

Flagella are too small to be resolved by visible light although their presence can be determined by dark-ground illumination or phase-contrast microscopy. Their staining depends on the aggregation of solid material upon their surfaces, to increase their apparent size. These methods are of little or no cytological value, and are not entirely to be relied upon, even for information upon the presence or absence of flagella, or upon their

---

Fig. 9

CELL ENVELOPES IN *BACILLUS*

In 'rough' septate bacilli the bacillus is usually divided centrally by a complete cross cell wall and subdivided by developing cross-walls at varying stages. The latter may stain as cell walls or cell membranes (*i.e.* basophilic) according to the stage of development.

(1) *B. cereus*, cell walls by Hale's method. Demonstrating only mature cross-walls. ×3000.

(2) *B. megaterium*, cell walls by tannic-acid-violet. Cross-walls at various stages can be seen; the most mature appear double, possibly because of tannic-acid complex deposited on their faces. ×4500.

(3) *B. megaterium* stained by haematoxylin. Developing cross-walls show as dark bars; the cell wall does not stain. Such basophilic areas have frequently been mistaken for nuclei or portions of nuclear structures. ×4500.

(4) *B. megaterium* stained by acid-thionin. This dye is less specific for the nuclear structures than Giemsa and stains also the basophilic areas of developing cross-walls, which appear either as readily recognisable bars or else as dots between the nuclei. In the latter form they have been confused with mitotic centrioles. ×4500.

(5, 6) Cell walls and cell membranes stained simultaneously by the Victoria blue method of Robinow and Murray. [33] The bacilli (*B. megaterium*) are slightly plasmolysed and the cell membrane is retracted from the mature cross-walls. The developing cross-walls show as dark bars, apparently composed of, or surrounded by, the same material as the cell membrane. ×3000.

arrangement, as they have, in the past, given contradictory evidence upon these points. There is no substitute for the electron microscope in the study of flagella, and luckily it is easy to make preparations for the purpose, in which the arrangement is natural and undisturbed. This can be done by growing the bacteria to be examined on a collodion mem-

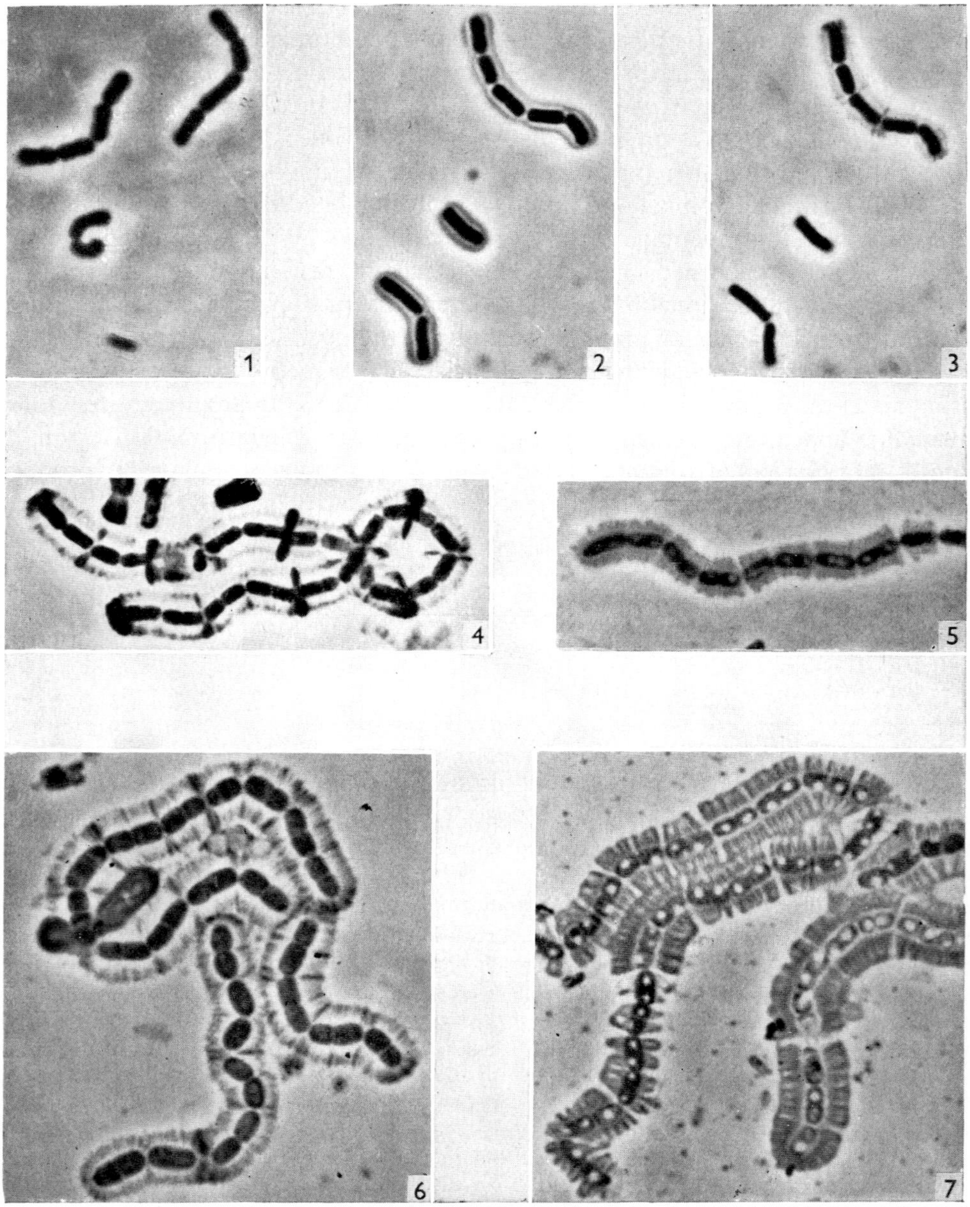

(Reproduced from the Journal of General Microbiology, by permission of Professor J. Tomcsik)

brane covering the surface of a solid culture medium [37] (Fig. 11).

ELECTRON MICROSCOPY

The electron microscope suffers from two defects in its application to biological materials; the specimen to be examined must be completely desiccated before introduction into the vacuum chamber, and the penetration of the electron beam is so low that only the thinnest specimens can be properly defined. For these reasons, until recently, the most valuable contribution of the electron microscope to bacterial cytology was in the study of flagella, but the perfection of techniques for embedding and ultra-sectioning have revolutionised the entire subject.

The scope of this section does not permit a detailed description of the techniques of electron microscopy, upon which complete books have been written, [38] but many examples are included among the illustrations in the following pages, of the types of information that can be obtained from this source.

Although ultra-sectioning now holds the centre of the stage, it is still far from being the only method available to electron microscopists. Some excellent observations have been made on crushed or disrupted material, for example, and have provided valuable information, not only upon the mode of attachment of flagella, but also upon the cell envelopes [39, 40, 41] (Fig. 12). Negative staining with heavy metals, cadmium, uranium, lead or phosphotungstic acid, has the effect of outlining raised structures in darker, opaque outline. It has revealed an unexpected complexity in bacterial outer surfaces [40, 42] (Fig. 13), and this has been borne out, in the case of endospores especially, by the technique of replica-casting [43] (Fig. 14). The latter method, which has also been applied with success to the elucidation of frozen-etched

---

Fig. 10

DEMONSTRATION OF CAPSULES BY TOMCSIK'S METHOD

These photomicrographs were made by Tomcsik's method for specific demonstration of antigenically active material by phase-contrast microscopy. The reaction of the antigen *in situ* with the antibody prepared against a chemically defined extract or suspension renders the structure in question visible to phase-contrast, and gives a simultaneous demonstration of its form and chemical composition. All plates are of a capsulated species of *Bacillus* ×2500.

(1) Capsulated bacteria without added antibody.

(2) The same, with the addition of an antibody active against the polypeptide fraction of the capsule.

(3) The same field as (2) with the further addition of an antibody active against the polysaccharide fraction.

It will be observed that the untreated capsule shows only as a diffuse, pale zone; after the addition of the polypeptide antibody it appears clearly defined but homogeneous, but the addition of the polysaccharide antibody reveals an unsuspected striated structure, with larger dark areas at the poles and points of division.

(4, 6) Originally non-capsulated bacteria which have developed a secondary capsule in a centrifuged deposit at room temperature. Polysaccharide antibody reveals exceptionally thick capsular septa, occasionally splitting in division.

(5, 7) The same as (4, 6) demonstrated by polypeptide antibody. In this case the polysaccharide septa appear as gaps in the capsule.

cell envelopes [44] (Fig. 17), requires further explanation. If a very thin layer of a heavy metal or silica or carbon, or a combination of such substances, is deposited on a surface, and then peeled away, the replica or cast will frequently reveal a high degree of accurate detail, in the electron microscope. Carbon can be employed very simply, by evaporation

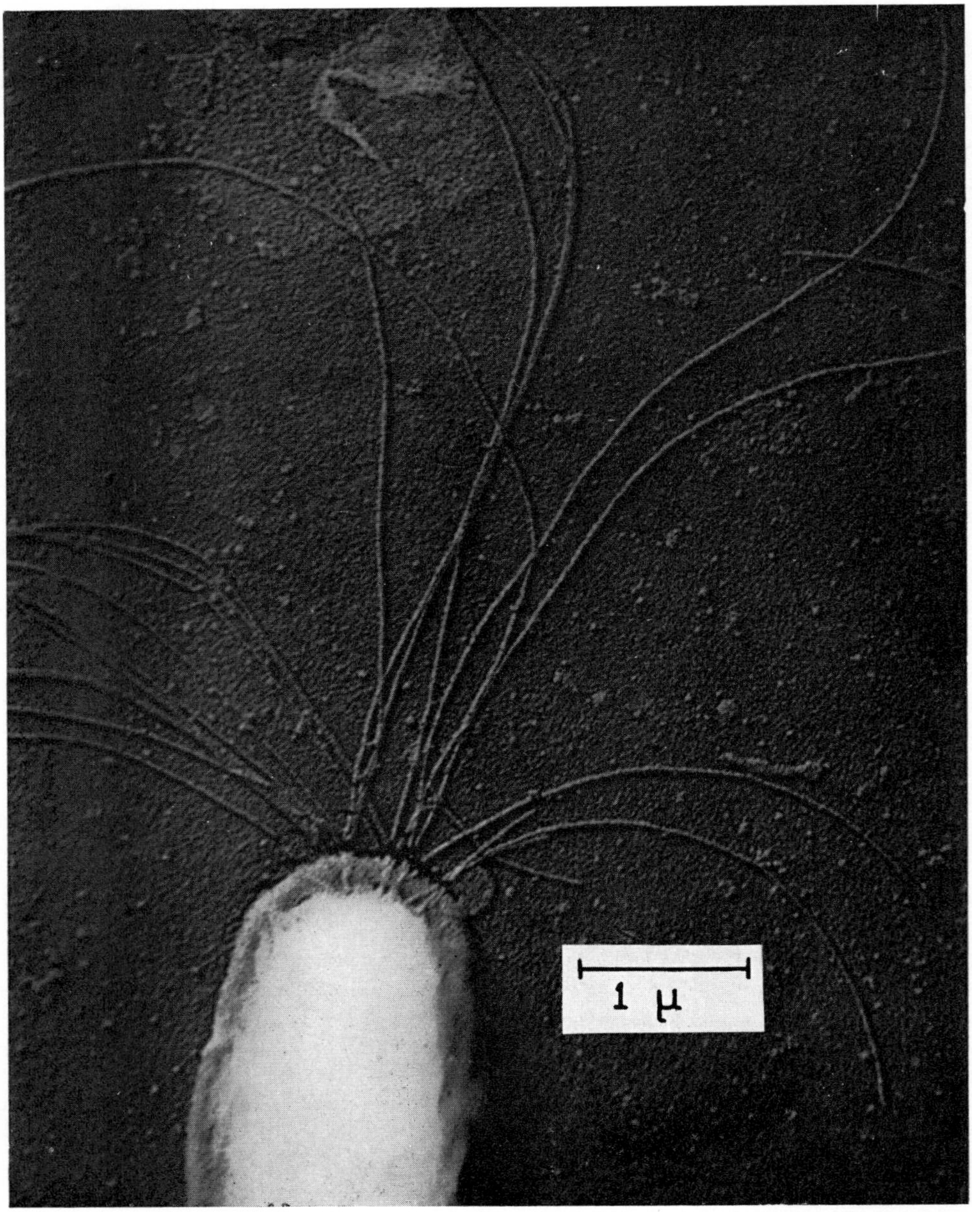

(Micrograph by Dr W. van Iterson. Reproduced from Biochimica et Biophysica Acta)

from a spark gap between graphite rods *in vacuo*.

Ultra-sectioning is, in principle, exceedingly straightforward, and does not differ in any essential detail from histological sectioning. Specimens are fixed, embedded and sectioned on a microtome, but as it is essential to produce sections that are transparent to the electron beam, the microtome must be very-low-geared, and the embedding medium capable of retaining its shape in slices that may be only a small fraction of a micron in thickness. Usually, setting plastics are employed, and are cut against the edge of a chip of plate-glass, instead of a steel knife. Polyesters and epoxy resins require that the specimens be previously dehydrated, exactly as for paraffin wax embedding, and this is true of most similar media. Water soluble media have been used upon occasion.

The most important single process in the preparation of biological material for sectioning is undoubtedly fixation, and in order to avoid coagulation of the softer structures, especially the nucleus, buffering of the fixative solutions is essential. (38, 45) The method called RK after Ryter and Kellenberger (46) uses calcium ions and amino-acids with an osmium fixative, but the addition of uranyl, which forms a salt-like compound with DNA, has been recommended as a modification. (45)

Positive staining, with the same electron dense compounds that are employed in negative staining, is frequently combined with ultra-sectioning. (47, 48)

PHASE-CONTRAST MICROSCOPY

The phase-contrast microscope has been applied to the study both of the bacterial nucleus and capsule. Early attempts to demonstrate the nucleus in living, unstained bacteria (49, 50) were not particularly helpful. The nuclei showed only as darker or lighter areas in approximately the positions indicated by stained preparations, but it is doubtful if, without the latter, they could have been interpreted. Later, Pulvertaft (51) demonstrated the reduction process in spore maturation (Fig. 62), but the lack of clear contrast between nuclei and cytoplasm was overcome only when Ross and his collaborators (52) devised the technique of mounting the material to be examined in media of various refractive indices, so as to match to the background that part of the structure which it is desired to suppress visually. By this means, they proved, fairly conclusively, that the water content of the endospore is reduced (previously a point of disagreement), and Mason and Powelson (54) produced the first really convincing serial micrographs of dividing nuclei in bacteria. Their method was to match the refractive index by varying the concentration of gelatine in the mountant, which also served as the nutrient medium. The nuclei appeared clearly in dark contrast to the cytoplasm.

An equally valuable method was discovered by Tomcsik, (55, 56) who showed that when antibodies against chemically defined

---

Fig. 11
BACTERIAL FLAGELLA

Spirillum serpens. Gold-shadowed electron micrograph, showing flagella passing through the cell wall to the protoplast.

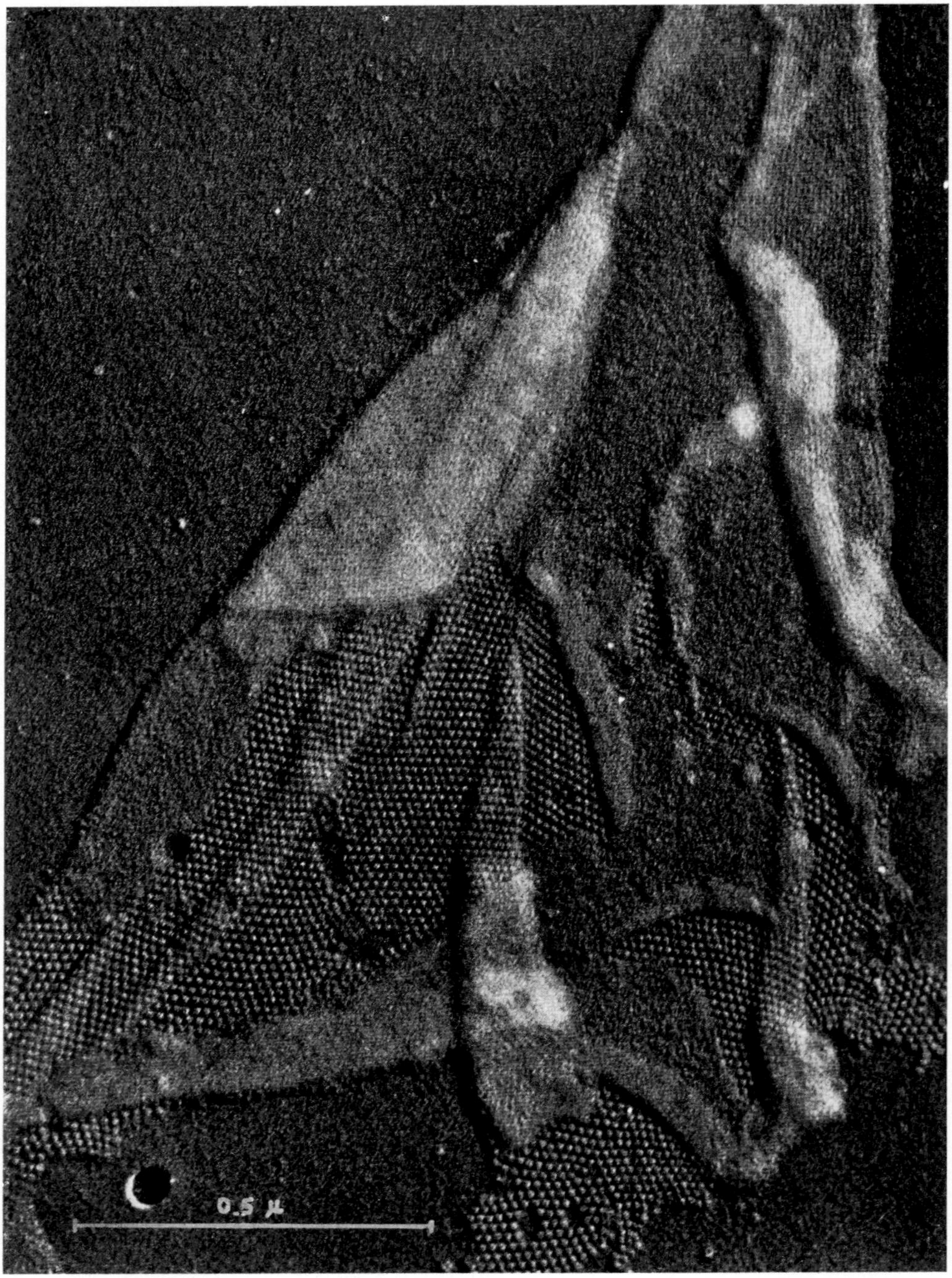

(Reproduced from Biochimica et Biophysica Acta, by permission of Dr A. L. Houwink)

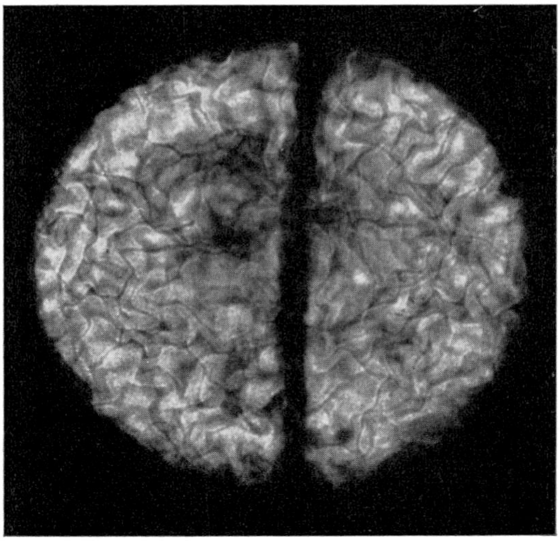

(*Electron micrograph by Dr H. A. Bladen*)

FIG. 13
COMPLEX CELL SURFACE

Cell wall of *Veillonella*, electron micrograph negatively stained by phosphotungstic acid, showing the deeply convoluted surface. ×70,000.

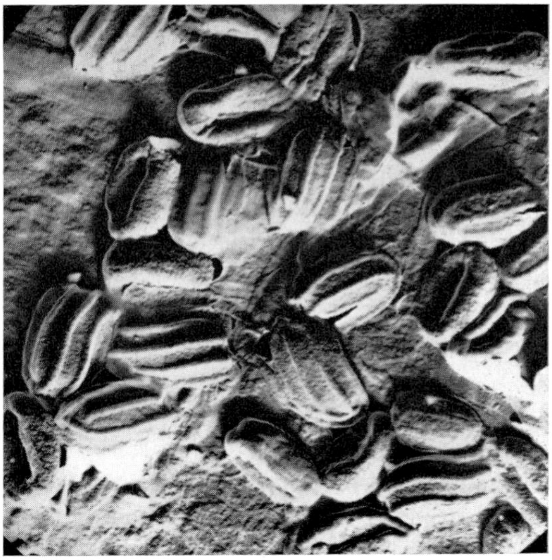

(*Reproduced from the Journal of General Microbiology, by permission of Dr D. E. Bradley*)

FIG. 14
SPORE SURFACE

Electron micrograph of *Bacillus polymyxa* spores. Carbon replica casts, showing the deeply sculptured surface. ×9,000.

fractions of, for example, the bacterial capsule are allowed to react with bacteria that contain the appropriate chemical fraction, in the field of the phase-contrast microscope, the antigen-antibody combination is made visible; presumably because the coagulation of the antigen changes its refractive index. By this means the capsule of certain Bacillus species has been demonstrated to consist of narrow lamina of polysaccharide and polypeptide elements, with larger masses of polysaccharide at the poles of the bacilli, and at the cross-walls (Fig. 10). The cell wall can also be made visible when it reacts with the homologous antibody. Non-specific proteins will enter into a salt-like combination with capsular components at appropriate values of pH, rendering them visible by phase-contrast microscopy. These methods can be

---

FIG. 12
ELECTRON MICROSCOPY OF CELL ENVELOPE MATERIAL

Spirillum sp., wall of crushed cell. The outer layer is represented by the striated pattern in the upper part of the plate, the inner by the pattern of regular globules below. ×100,000. Compare Figures 13 and 17.

used with even greater success if the structures are separated by partial digestion with such enzymes as lysozyme or trypsin.

The further development of these techniques, by the use of antibodies specific for cytoplasmic and nuclear components, offers a most promising field for study in the cytology of bacteria and other cells.

COLONY PREPARATIONS

Although a bacterial surface colony upon solid medium is a semi-artificial formation, the study of its minute structure is often of biological interest and diagnostic importance. It is capable of providing evidence of the natural relationship of the bacteria to one another, and, especially in the study of dissociation, may indicate differences in structure and behaviour which are not always obvious by other methods of examination.

Entire colonies may be embedded and sectioned, like portions of tissue, but as the colonies are usually exceedingly thin and flat (much more so than they appear to the unaided eye), whole mounts can be made upon slides or coverslips. These are usually termed impression preparations; [57] they are best made from very small colonies, although quite large ones can be mounted if the growth is sufficiently tough.

A small piece of medium bearing the desired colonies, is cut out with the point of a knife and placed, face downward, upon a slide or coverslip. Surface tension will suffice to keep the medium, and the attached colonies, firmly pressed to the glass. It is then fixed, in its entirety, preferably in Bouin's solution, until the medium is blanched throughout, and can be peeled away from the glass, leaving the colonies adhering to the surface of the coverslip. The preparations may then be

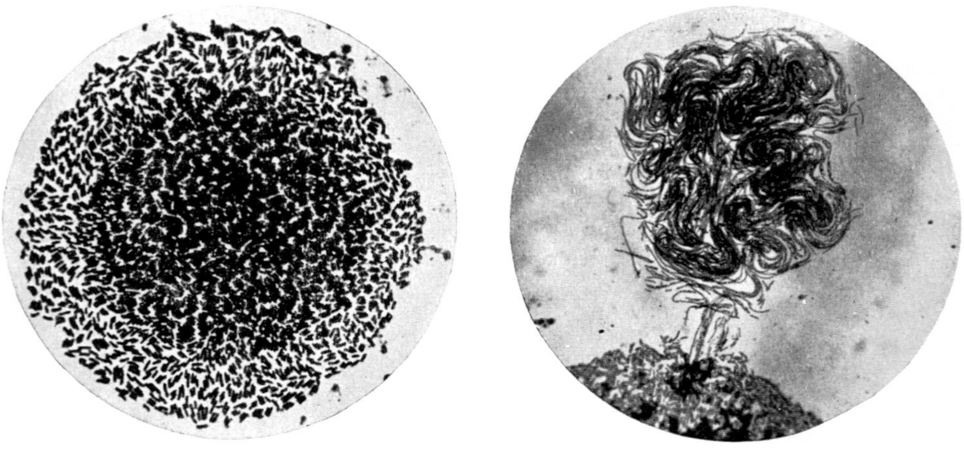

A B

(*Reproduced from the Journal of Pathology and Bacteriology*)

FIG. 15
SMOOTH AND ROUGH COLONIES

A. Smooth colony, *E. coli*, impression preparation ×500.

B. Rough colony arising from the perimeter of a smooth colony, *Shigella flexneri*, ×300.

washed, stained and mounted. Sometimes they are of great beauty and surprising complexity, even when very tiny (Fig. 15). The best medium for this purpose is blood agar, it is very adhesive when fixed, and becomes firmly attached to the glass, so that it can be peeled away without danger of sliding the medium laterally and destroying the colony. Plates should be inoculated with the rounded tip of a glass rod, to avoid scratching the surface, they should be perfectly dry, and free from bubbles and other irregularities.

The technique of growing bacteria on a collodion film for electron microscopy has already been mentioned in reference to bacterial flagella, but it is also possible, by this method, to examine whole microcolonies, either of bacteria [(58)] or of their L-forms. [(59)] A drop of collodion solution is floated on water over an agar plate. The water is pipetted away, and the resulting film inoculated. When the culture has grown for a few hours, the film bearing the growth is cut away, floated off the agar, and mounted on a grid, for examination under the electron microscope.[(37)]

REFERENCES

1. Stille, B. (1937). *Arch. Mikrobiol.* **8**, 125.
2. Piekarski, G. (1937). *Arch. Mikrobiol.* **8**, 428.
3. Neumann, F. (1941). *Zentbl. Bakt. ParasitKde, I, Orig.* **103**, 385.
4. Robinow, C. F. (1942). *Proc. R. Soc.* **B**, **130**, 299.
5. Robinow, C. F. (1944). *J. Hyg., Camb.* **43**, 413.
6. Paillot, A. (1919). *Annls Inst. Pasteur, Paris*, **33**, 403.
7. Gutstein, M. (1924). *Zentbl. Bakt. ParasitKde, I, Orig.* **93**, 393.
8. Vendrely, R. & Lipardy, J. (1946). *C. r. Séanc. Acad. Sci. Paris*, **223**, 342.
9. Robinow, C. F. (1956). *Symp. Soc. gen. Microbiol*, **6**, 181.
10. Cassel, W. A. (1950). *J. Bact.* **59**, 185.
11. Bisset, K. A. (1954). *J. Bact.* **67**, 41.
12. DeLamater, E. D. (1951). *Stain Technol.* **26**, 199.
13. Klieneberger-Nobel, E. (1947). *J. gen. Microbiol.* **1**, 1.
14. Grace, J. B. (1951). *J. gen. Microbiol.* **5**, 519.
15. Badian, J. (1933). *Arch. Mikrobiol.* **4**, 409.
16. Bisset, K. A. (1948). *J. Hyg., Camb.* **46**, 264.
17. Nakanishi, K. (1900). *Münch. med. Wschr.* **47**, 187.
18. Stoughton, R. H. (1929). *Proc. R. Soc.* **B**, **105**, 469.
19. Stoughton, R. H. (1932). *Proc. R. Soc.* **B**, **111**, 46.
20. Allen, L. A., Appleby, J. C. & Wolf, J. (1939). *Zentbl. Bakt. ParasitKde*, **II**, **100**, 3.
21. Bisset, K. A. & Hale, C. M. F. (1951). *J. Hyg., Camb.* **49**, 201.
22. Robinow, C. F. (1953). *J. Bact.* **65**, 378.
23. Davis, J. C. & Mudd, S. (1957). *J. Histochem. Cytochem.* **5**, 254.
24. Niklowitz, W. (1958). *Zentbl. Bakt. ParasitKde, I, Orig.* **173**, 12.
25. Weibull, C. (1953). *J. Bact.* **66**, 137.
26. Smith, I. W., Wilkinson, J. F. & Duguid, J. P. (1954). *J. Bact.* **68**, 450.
27. Duguid, J. P., Smith, I. W. & Wilkinson, J. F. (1954). *J. Path. Bact.* **67**, 289.
28. Tomcsik, J. & Grace, J. B. (1955). *J. gen. Microbiol.* **13**, 105.
29. Murray, R. G. E. & Robinow, C. F. (1952). *J. Bact.* **63**, 298.
30. Weibull, C. (1955). *Expl Cell Res.* **9**, 139.
31. Hale, C. M. F. (1953). *Lab. Pract.* **2**, 115.
32. Finkelstein, H. & Bartholomew, J. W. (1958). *Stain Technol.* **33**, 177.
33. Robinow, C. F. & Murray, R. G. E. (1953). *Expl Cell Res.* **4**, 390.
34. Burdon, K. L. (1946). *J. Bact.* **52**, 665.
35. Jarvi, O. & Levanto, A. (1950). *Acta path. microbiol. Scand.* **27**, 473.
36. Hale, C. M. F. & Bisset, K. A. (1958). *J. gen. Microbiol.* **18**, 688.

37. Hillier, J., Knaysi, G. & Baker, R. F. (1948). *J. Bact.* **56**, 569.
38. Kay, D. H. (1965). *Techniques for Electron Microscopy*. 2nd ed. Oxford: Blackwell.
39. Houwink, A. L. (1953). *Biochim. Biophys. Acta*, **10**, 360.
40. Salton, M. R. J. & Williams, R. C. (1954). *Biochim. Biophys. Acta*, **14**, 455.
41. Abram, D., Vatter, A. E. & Koffler, H. (1966). *J. Bact.* **91**, 2045.
42. Bladen, H. A. & Mergenhagen, S. E. (1964)· *J. Bact.* **88**, 1482.
43. Bradley, D. E. & Williams, D. J. (1957). *J. gen. Microbiol.* **17**, 75.
44. Remsen, C. & Lundgren. D. G. (1966). *J. Bact.* **92**, 1765.
45. Fuhs, G. W. (1965). *Bact. Rev.* **29**, 277.
46. Ryter, A. & Kellenberger, E. (1958). *J. Ultrastruct. Res.* **2**, 200.
47. Murray, R. G. E., Steed, P. & Elson, H. E. (1965). *Can. J. Microbiol.* **11**, 547.
48. Ellar, D. J., Lundgren, D. G. & Slepecky, R. A. (1967). *J. Bact.* **94**, 1189.
49. Tulasne, R. (1949). *C. r. hebd. Séanc. Acad. Sci., Paris*, **229**, 561.
50. Stempen, H. (1950). *J. Bact.* **60**, 81.
51. Pulvertaft, R. J. V. (1950). *J. gen. Microbiol.* **4**, 14.
52. Barer, R. Ross, K. F. A. & Traczyk, S. (1953). *Nature, Lond.* **171**, 720.
53. Ross, K. F. A. & Billing, E. (1957). *J. gen. Microbiol.* **16**, 418.
54. Mason, D. J. & Powelson, D. M. (1956). *J. Bact.* **71**, 474.
55. Tomcsik, J. & Guex-Holzer, S. (1954). *J. gen. Microbiol.* **10**, 97.
56. Tomcsik, J. & Guex-Holzer, S. (1954). *J. gen. Microbiol.* **10**, 317.
57. Bisset, K. A. (1938). *J. Path. Bact*, **47**, 223.
58. Bisset, K. A. & Pease, P. E. (1957). *J. gen. Microbiol.* **16**, 382.
59. Pease, P. E. (1957). *J. gen. Microbiol.* **17**, 64.

CHAPTER 3

Surface Structures

THE CELL ENVELOPES

Bacteria are of such small size that the adoption, in a fluid medium, of any other form than that of a sphere, argues considerable rigidity of structure. Were this rigidity absent, the forces of surface tension, relatively enormous in such a case, would force the bacterium to adopt the form possessing the smallest proportion of surface area to volume. While it is true that some bacteria are almost perfect spheres, although these are rather fewer than is often supposed, the majority are rod-shaped, usually slightly spiral, [1] and sometimes more markedly spiral so that this feature is sufficiently obvious to attract attention. Their rigidity is further emphasised by the absence of flexion in their movements, except in certain specialised forms such as myxobacteria and spirochaetes. Enforced flexion causes fracture and distortion of the bacterium. This rigidity is due to the possession of a cell wall of great strength.

An understanding of the nature and behaviour of the cell wall and membranes of bacteria is a necessary preliminary to studies of all kinds in bacterial cytology.

The most usual and indeed the most fundamental error which arises from such neglect is the assumption, still frequently made, that bacteria are normally unicellular, whereas in very many groups, ranging in morphology from cocci to branched filaments, a single bacterium may contain from two to a dozen relatively tiny cells (Figs. 7, 8 & 9). The partitions between these cells may break down in the course of the autogamous processes that accompany sporulation, and at other times, but are usually found in the vegetative cells.

The cell wall is permeable. It does not grow, but is secreted, in certain well-marked areas [2, 3] (Fig. 16).

The cell wall is difficult to demonstrate and is seldom observed in preparations stained by the usual methods of routine bacteriology. The cross-walls are laid down, in the dividing cell, by cytoplasmic septa and associated bodies such as mesosomes. These complexes usually stain more readily than any other structures in the bacteria. [4] When the multicellular nature of such bacteria goes unrecognised, as it frequently does, these basophilic septa are liable to be confused with nuclei or cytoplasmic inclusions [5] (Fig. 2).

The complex cellular structure possessed by many bacteria has long been on record, although seldom adequately appreciated. Such multicellularity, sometimes with as many as twenty cells in each bacterium, separated by cross-walls and cytoplasmic septa, has been demonstrated in nearly all Gram-positive bacteria. [6] A greater degree is found in the giant bacteria such as Caryophanon and Oscillospira, [7] and even cocci can have complex internal cross-walls [6, 8] (Figs 3 & 8).

That the multicellularity of these bacteria is fundamental, and by no means a superficial subdivision of filaments by the irregular growth of septa, is shown by the observation of Tomcsik, [9] that in *Bacillus anthracis* the characteristic division of the bacillus into four small cells extends also to the polypeptide capsule, in which lines of demarcation can be seen, corresponding to the positions of the cross-walls internally (Fig. 10).

The chemical composition of bacterial cell walls has been much studied of recent years, but some of the earlier findings were not strictly-speaking cytology, since they were not correlated with any scheme of physical structure. Accurate observations may be said to have commenced with Holdsworth [10, 11] who isolated a protein-carbohydrate complex from the cell wall of a corynebacterium. Emphasis has been laid upon the value of cell wall composition as a diagnostic character in systematology, and especially upon differences between those of Gram-positive and Gram-negative bacteria. The former have a relatively small number of distinguishable amino-sugars in their cell walls, [12] and the various genera are characterised by the possession of specific amino-sugars, [13] whereas Gram-negatives have fewer amino-sugars, and these are rarely characteristic. However, it has been pointed out that the differences may not be fundamental, since the Gram-negative pattern is suggestive of blocked synthesis, [14] and corresponds to the well-known observation that bacterial variants can lose their Gram-staining under adverse conditions.

A considerable amount of information has been obtained by electron microscopy. The external surfaces of the cell envelopes appear to be composed of ordered aggregates of macromolecules (Fig. 12) that can be revealed by metallic shadowing [15, 16, 17] or by the carbon-replica technique. [17, 18] The arrangement of layers has been studied by ultra-sectioning [19, 20, 21, 22] and by the freeze-etching method. [18, 23] The last is effected by cutting or scraping the surface of a frozen block of bacteria, and making a carbon replica of the edges of the broken envelopes; it is capable of showing the arrangement of overlapping membranes in some detail (Fig. 17).

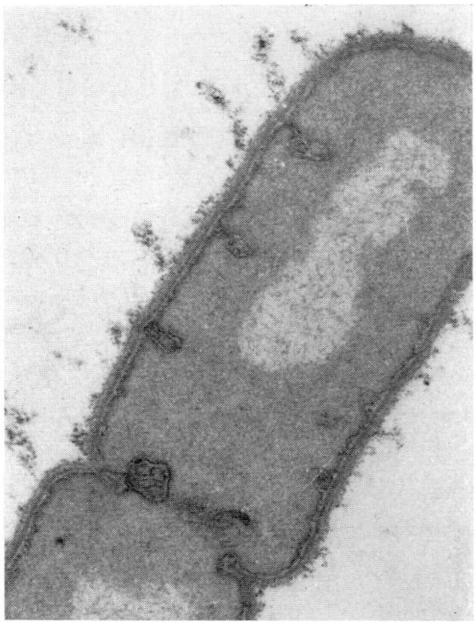

(Reproduced from the Journal of Applied Bacteriology, by permission of Dr P. D. Walker)

FIG. 16
GROWING POINTS

Electron micrograph of a section of *Bacillus stearothermophilus*, showing a row of mesosomes, representing points of growth and incipient division of the cell envelopes. × 12,000.

As already suggested, a wide variation is found in the chemical composition of bacterial cell envelopes.[12] Remsen and Lundgren [23] described a three layered complex in Ferrobacillus, each layer being 100 Å in thickness. The outermost was of lipoprotein-lipopolysaccharide; the middle layer, globular protein attached to fibrillar mucopeptide; the inner was the semipermeable membrane of the cell, which will be discussed below. Nermut [24] suggested that the basic type of Gram-positive wall structure is found in *Bacillus megaterium*, and consists of a rigid layer of mucopeptide, about 100 Å thick, in which the fibrillar structure is lengthwise. This is surmounted by a plastic layer of mucopolysaccharide (teichoic acid), with the fibrils at right-angles to the wall. The two schemes are essentially similar fundamentally. Remsen and Lundgren showed bristle-like fibrils of mucopolymer (mucopolysaccharide) internal to their central layer, and Murray and his collaborators, [16, 20, 25] who claimed a more complex structure, with four layers, in *Bacillus polymyxa* and in Gram-negative bacteria, external to the cell membrane, also indicated the presence of mucopeptide between wall and membrane. The latter workers emphasised the importance of the mucopeptide in the structural rigidity of the wall, and suggested that the membrane and mucopeptide alone might provide an irreducible minimum of envelopes, for proper functioning where the remainder of the wall is absent or has been removed, as, for example, in L-forms, protoplasts and mycoplasmas. The very thin, but rigid wall of Bacillus swarm cells may be of this type [26] (Fig. 84). The composition of cell walls is more fully discussed by Rogers and Perkins. [27]

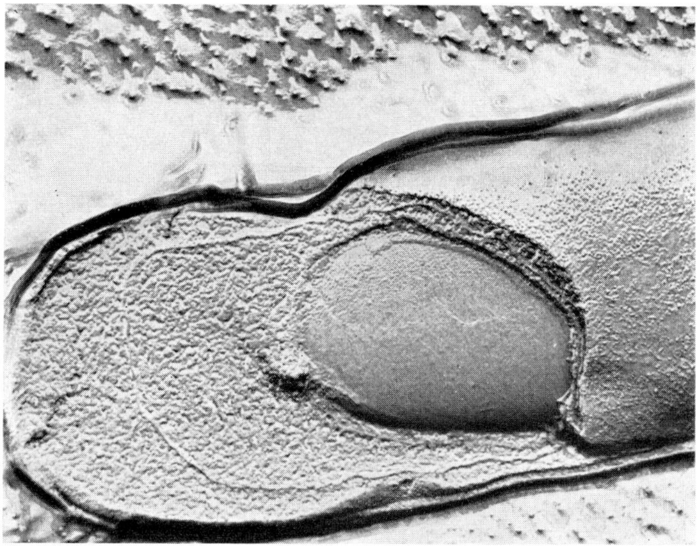

(*Electron micrograph by Dr C. C. Remsen*)

FIG. 17
FREEZE-ETCHING

Platinum-carbon replica of frozen-etched surface of a sporulating *Bacillus cereus*. Several layers of envelopes can be discerned by their broken edges; notably the fore-spore membrane in the upper part of the picture. Below the spore can be seen micellar projections from the cell membrane. ×45,000.

The cell membrane or cytoplasmic membrane, as it is variously called, can be demonstrated by light microscopy in the intact cells of some (but not all) bacteria, by selective staining of slightly plasmolysed material [28] (Fig. 9), and rather elegantly by a fluorescent derivative of the antibiotic polymyxin, which appears to attack it selectively. [29] The visualisation of the cell-membrane as a separate structure has been achieved in lysozyme digested bacteria by Weibull [30, 31] and Tomcsik and Guex-Holzer. [9, 32] In the last stages of dissolution, the protoplasts appeared as 'ghosts', and the empty membrane survived momentarily. The behaviour of the cell membrane at cell division provides additional evidence that it is a positive structure, and not merely an interface. The septa which initiate cell division are clearly resoluble, although surrounded on both sides by cytoplasm, [20, 33] and can be demonstrated in sections (Fig. 18).

The large quantity of basophilic material contained in, or associated with the cell membrane may be a cause of confusion (Fig. 9). Unlike the cell wall, it is easily stained by almost any method. This applies especially to the cytoplasmic septa and the mesosomes

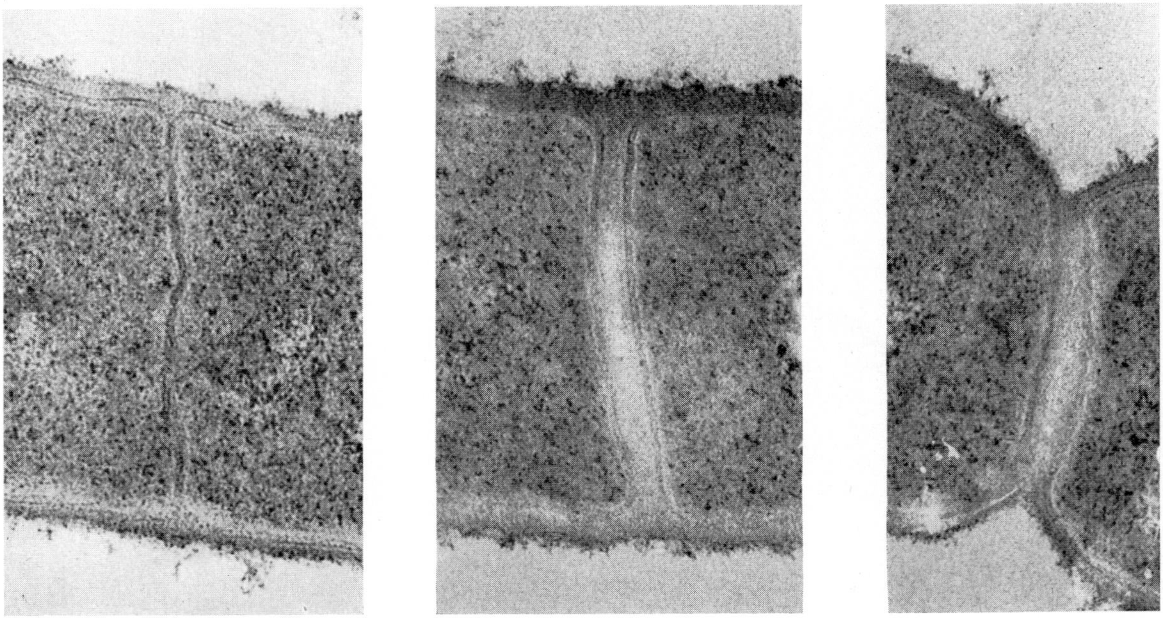

a *b* *c*

(*Electron micrographs by Dr Roger Cole*)

FIG. 18
CELL FISSION

Three stages in cell fission of *Bacillus subtilis*, seen in sections. *a*, a very fine septum, derived from the cell membrane, divides the daughter cells; *b*, cell wall material is formed between two layers of membrane; *c*, fission proceeds by constrictive division of the septum. $\times 60{,}000$.

and growing points, which are associated with the membrane. These structures will stain well with many dyes supposedly specific for chromatin, mitochondria, reticulocytes, etc. It is interesting to consider the observation of Pijper [34] that somatic agglutination occurs by the adherence of the bacteria at the tips of the cells. Not only is the cell wall very thin at this point, being in the process of formation, but one of the main aggregations of stainable and cytochemically active material in the cell membrane occurs immediately underneath, so that it would appear that the growing points, and presumably also the almost identical cytoplasmic septa, both with their associated mesosomes, may be a major somatic antigen.

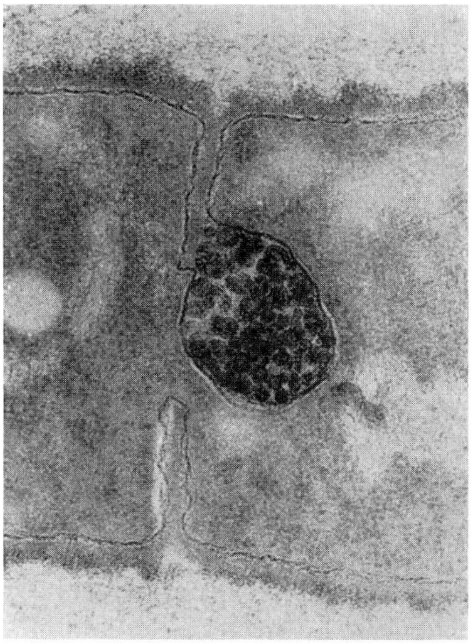

(Reproduced from the Journal of Bacteriology, by permission of Dr D. G. Lundgren)

FIG. 19
MESOSOME

Electron micrograph of a section of *Bacillus megaterium* showing a mesosome preceding the ingrowth of the membrane at cell division. ×15,000. Compare Figure 18.

The cell membrane is largely responsible for the ease with which entire bacteria within the cell wall can be stained, and when this stainable cortex is viewed through the tips of the cell, the well-known optical illusion of 'bipolar staining' is observable. Ribose nucleic acids in the cell envelopes have been claimed to cause the phenomenon of Gram-positivity, where it occurs. [35, 36] It has also been proposed [37] that Gram-positivity is due to the presence of a phosphoric ester, the occurrence of which is independent of the concentration of ribose nucleic acid. Or that a difference in the tyrosine content of the pentose nucleoproteins of Gram-positive and Gram-negative bacteria might account for the difference.[38]

Probably one of the main functions of the powerful mordanting agents, such as tannic acid [39] or phosphomolybolic acid, [40] that serve to render the cell wall of bacteria stainable, and also distinguishable from the cytoplasm, is the coagulation of the basophilic material in the cell membrane (Chap. 2).

The semi-permeable membrane can be regarded as the fundamental structure, the possession of which is the characteristic of a cell. Fitz-James [41] has suggested that the development of a lipid micelle layer, upon which proteins and nucleic acids could function, was the early, decisive step in the evolution of life as we know it. Membranes are believed to consist of a layer of lipid between two layers of protein, and although bacterial membranes have a lower content of sterols and certain phospholipids than do those of most other cells, [42] this sandwich structure can be made out in at least some ultrasections. [41, 43]

Membranes of a specialised type, with an active basophilic component are found in the mother-cells of Rhizobium swarmers. [44] These resemble the polysaccharide-complex cross-walls of Bacillus species and are lined on both sides with basophilic material resembling the normal, membranous septa, but greatly thickened. They appear to provide a secretory lining to the lumen of the cells wherein the tiny swarmers are formed. The misconceptions which have arisen from the appearance of these 'barred cells' are discussed in a later section (Chap. 7).

The fundamentals of bacterial cell division described by Schaudinn in 1902, 1903 (Chap. 1) have been well substantiated by later work. [20, 33, 45] The two main morphological types of bacteria, which correspond to the 'smooth' and 'rough' colony forms, divide by constriction of the cell wall and by the formation of cross-walls respectively. In the 'rough' type, which is especially typical of the large, Gram-positive bacilli, the division of the cells by cross-walls proceeds more rapidly than does their complete separation, so that the coiled bands of filaments are

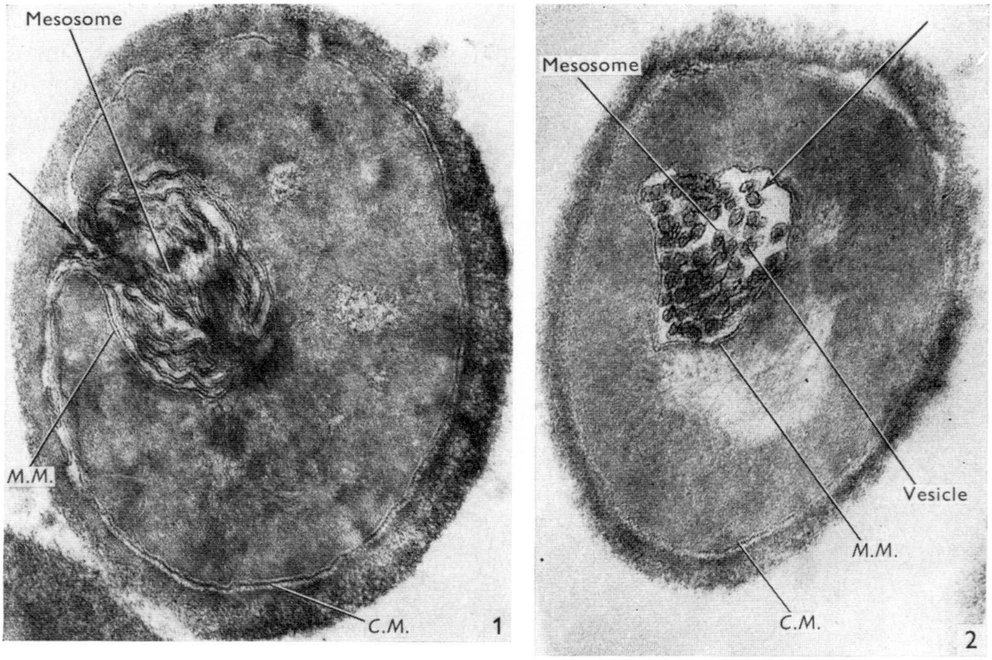

(Reproduced from the Journal of General Microbiology, by permission of Dr C. D. Beaton)

FIG. 20
STRUCTURE OF MESOSOMES

Electron micrographs of sections, showing mesosomes. C.M., cell membrane; M.M., mesosome membrane. Arrow shows the point of attachment of cell envelopes and mesosome. One interpretation of these micrographs is that the mesosomes have been sectioned in two directions, at right-angles. Beaton (56) considers that two types of mesosome are shown. ×160,000.

formed that give the well-known 'Medusa-head' colony appearance. [46] The actual separation of rough bacilli does not necessarily occur immediately after the completion of cell division, in any case (Fig. 18).

It has long been recognised that the growth of the cell envelopes at the poles of the cells (*i.e.* at the poles of bacilli, and also at the points where new cross-walls are being formed) is usually marked by concentrations of cytochemically active material, [2, 4, 47, 48] the growing points. These are closely associated with the cell membrane, and recent studies made by means of ultra-sectioning and electron microscopy have shown that an important element in their composition is the mesosome. This structure (Figs 16, 19, 20 & 21), is essentially an invagination of the membrane [21, 41, 49, 50] forming a discrete, convoluted body. It is a site of oxidation-reduction activity [51, 52] and has, not only a role in cell-wall synthesis [41, 53, 54] but also in the division of the nucleus. [41, 47, 55] It is to be presumed that it represents a device for bringing the maximum surface of membrane into action at one point, and that its main function is the transfer of energy. Mesosomes have been demonstrated in a wide variety of bacteria [50, 56, 57] but are especially obvious in sporing bacilli. [21, 54, 55]

Cell division of both types commences with an annular ingrowth of the membrane, which may or may not be preceded by a mesosome at the incipient point of division (Figs 16 & 18). The ingrowth of the cell wall follows that of the cell membrane. It is at these points that the main growth of the cell envelopes occurs; [4, 5] the new surface being, as it were, passed inwards and outwards around the edge of this division between the cells, forming both a septum or constriction internally, and a new portion of cell wall externally, whereby the growth of the bacillus proceeds (Figs 18 & 22). There may be a single point of division, or several of them [58] according to the morphology of the bacillus. All Gram-positive and many Gram-negative bacteria have a septate structure [4, 5, 6] and thus, numerous potential growing-points. In studying this, Tomcsik [9, 32] and Cole [3, 58] both employed immunological labelling, visualised by phase-contrast and by fluorescence microscopy respectively. The former showed that the capsule also develops mainly from the junctions of cell wall and cross-walls, the latter distinguished between new and previously existing cell wall material, and demonstrated the growth of the wall from the same areas. Both methods reveal a degree of complexity in the cell envelopes, even when division is not complete, that can with difficulty be seen by other methods. Some electron micrographs of ultra-sections show mesosomes representing new growing-points, developing simultaneously along the length of a bacillus [49, 59, 60] (Fig. 16), but these are sporing bacilli, of a septate type. The large number of growing-points shown by immunofluorescence is rather surprising, and it must be remembered that subjection of a growing culture to the action of antibodies against the cell envelopes is well known to produce S→R variation, and the R variants are much more filamentous and septate than the original S. [46]

The dictum that the main growth of the cell envelopes takes place at the poles or future poles of the *cells* was explained in the previous edition of this book and elsewhere, [4, 5] but it has been widely misconstrued as meaning that all bacteria grow from the poles of the bacillus. [12, 59] It is, however, true that unicellular bacteria may grow from a single growing-point, at or near

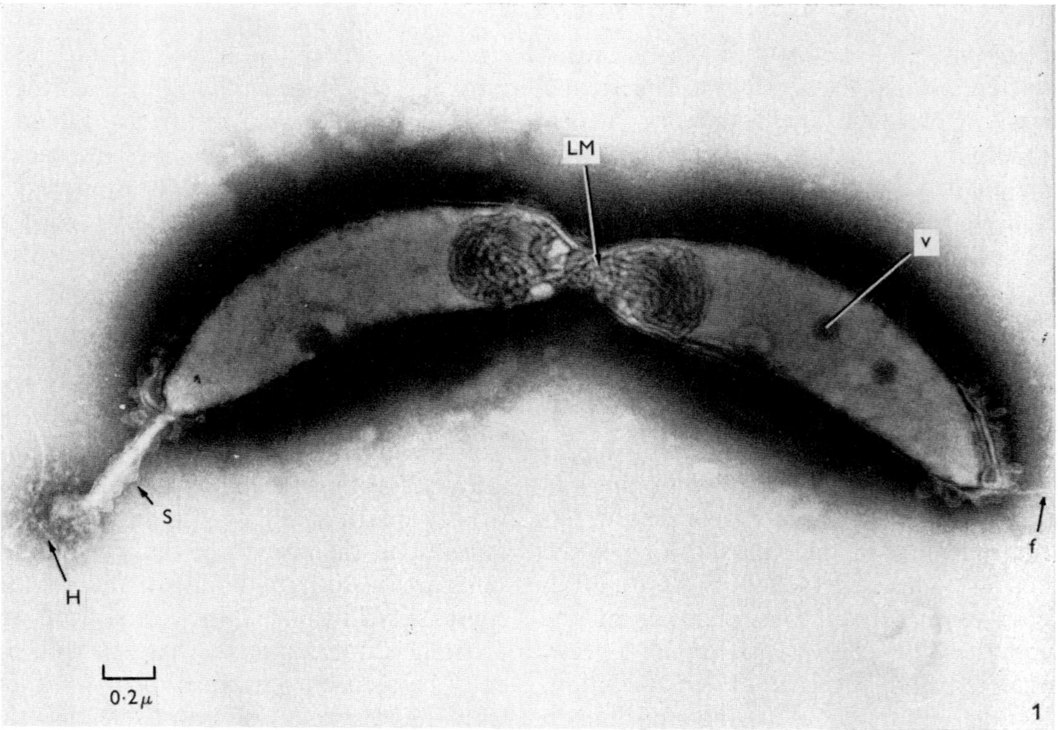

1

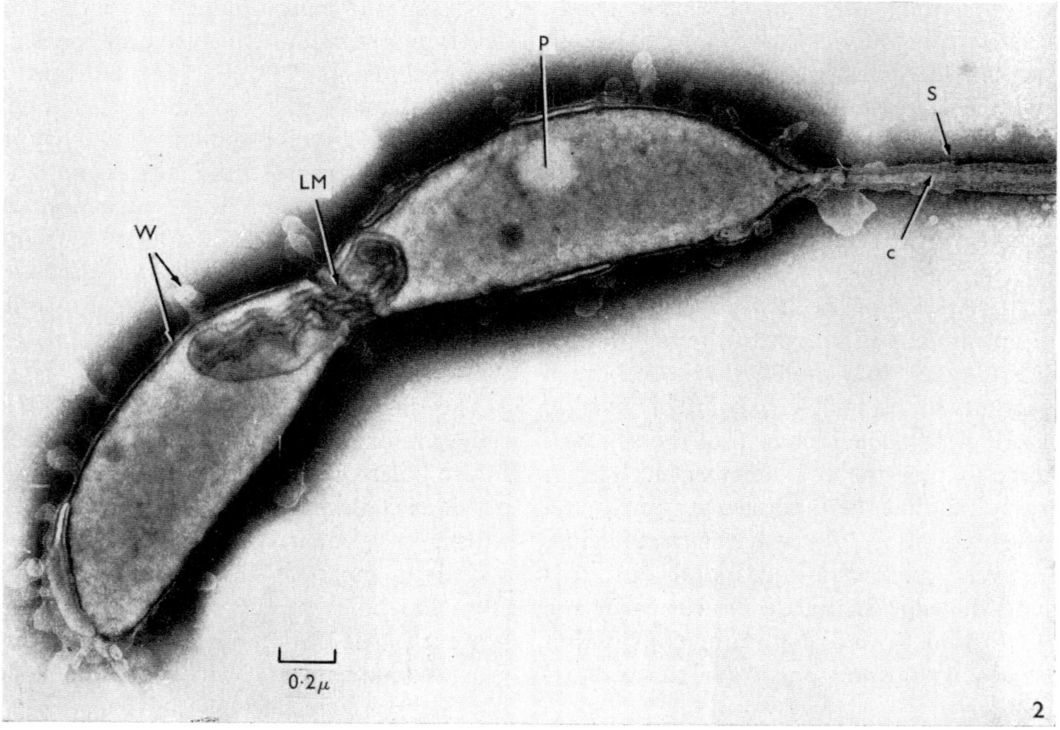

2

(*Reproduced from the Journal of General Microbiology by permission of Dr Germaine Cohen-Bazire*)

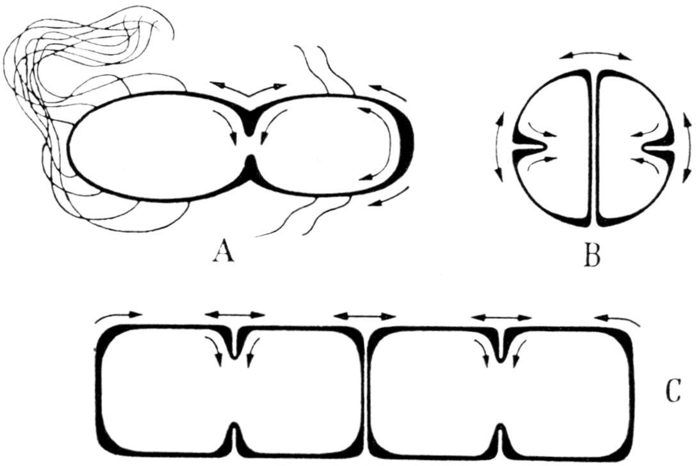

FIG. 22
GROWTH OF CELL ENVELOPES

A, unicellular bacterium, growing from tip of daughter cell and point of fission; B, C, septate coccus and bacillus, growing from points of junction of cell wall and cross-walls. This diagram is of some historical interest. Originally published in 1957, it shows conclusions exactly in accordance with information later derived from ultra-sectioning, immunofluorescence etc. Figure by the author.

the pole; [2] but even these grow also from the point of division, when they are dividing, as they usually are. The point of division then becomes the growing-point of one or both daughter cells, so that there is no fundamental difference between the growing-points at the poles, and those at the points of division (Fig. 22).

There is a considerable amount of evidence to suggest that unicellular bacteria grow mainly, or more rapidly, towards one pole, and that they do not truly divide into two equal daughter cells, but rather produce a bud, of size equal or almost equal to the mother cell. This evidence is derived from several sources. When treated with neutral formaldehyde, young cultures retain their basophilia, whereas older ones tend to lose it. In dividing bacilli and cocci, usually one of each dividing pair appears older, and one younger. [61, 62] Secondly, the behaviour of the flagella at cell division is markedly signi-

---

FIG. 21
MESOSOMES IN CAULOBACTER

Electron micrographs of sections through *Caulobacter crescentus*. LM, large mesosome; S, stalk; H, holdfast; V, granule; f, flagellum; p, granule; W, cell wall. ×36,000.

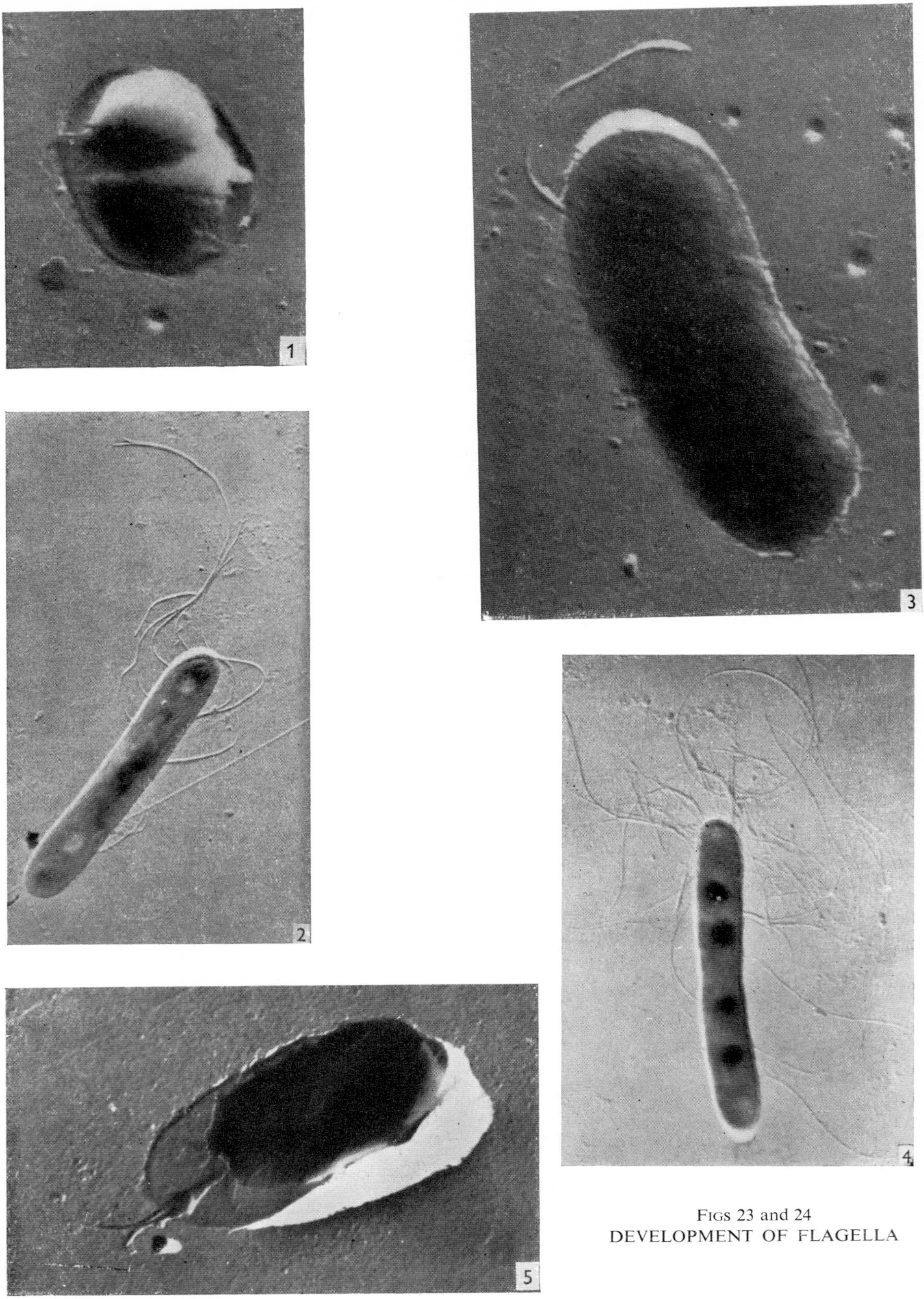

FIGS 23 and 24
DEVELOPMENT OF FLAGELLA

Development of flagella in the germinating microcyst. The resting cells of flagellated bacteria are devoid of flagella; on germination these develop first at the poles of the cell, especially at the pole remote from the growing point. Electron micrographs, gold-shadowed.

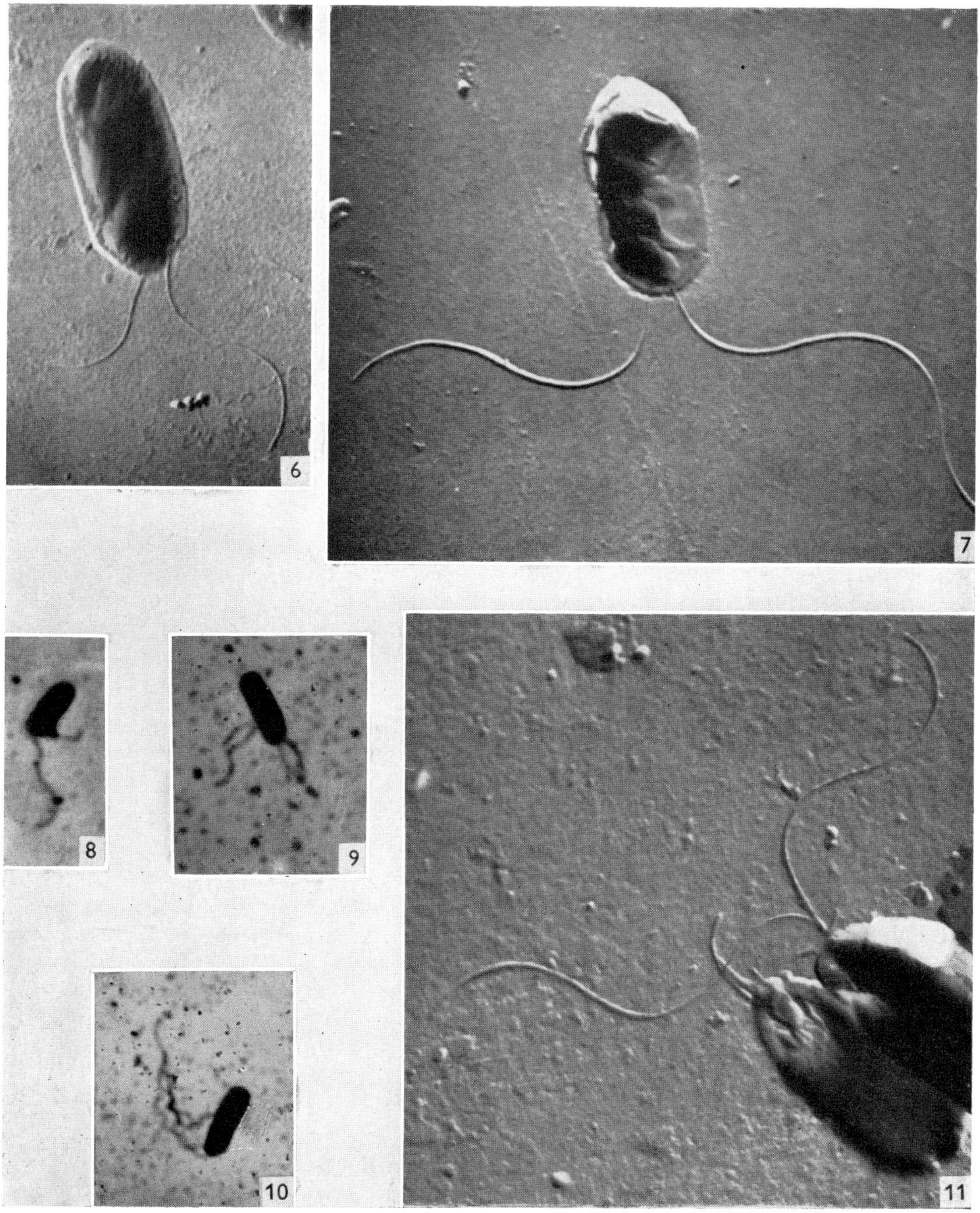

(1, 2, 3, 4) *Salmonella typhi*, stages in the germination of the microcyst.

(1) Microcyst, without flagella. ×30,000.

(2, 4) Young vegetative cells with short flagella concentrated towards one pole of the cell. ×9000 and ×7000.

(3) Germinating microcyst with very short flagella towards both poles. ×27,000.

(5, 6, 7) *E. coli*, stages in the germination of the microcyst.

(5) Germinating microcyst with two very short flagella originating at one pole of the cell from an obvious blepharoplast. ×20,000.

(6, 7) Two polar flagella further developed. ×16,000.

(8, 9, 10) Development of flagella demonstrated by silver impregnation stain. ×3000.

(8) *Proteus*, germinating microcyst.

(9, 10) *Sal. typhi*, very young cells with sub-polar flagella.

(11) Germinating microcysts of *Pseudomonas fluorescens*. The flagella emerge more closely together than in the case of the foregoing examples, which will eventually become peritrichous. Electron micrograph, gold-shadowed. ×16,000.

ficant, especially in unicellular (S phase, Gram-negative) bacteria. In electron micrographs, both of growing cultures and of germinating microcysts, the flagella appear progressively shorter towards one pole, where the wall is relatively thin and electron transparent. (2, 63) Frequently, one daughter cell has a full quota of flagella, whereas the other has few, or very short ones (2, 62, 64) (Figs 22, 23 & 24). This interpretation has been contradicted, on the evidence of stained preparations, (65) but it has been confirmed, not only by parallel electron microscopy of the same material, (64) but by observation of living bacteria (66) and by immunouranium labelling. (67) Thirdly, the existence of polar growing-points has been established by Bergersen (48) who showed that, under the influence of sub-lethal concentrations of chloramphenicol, bacteria developed typical basophilic concentrations in the cell membranes, from which irregular side-branches were produced. Direct photographic methods have produced evidence of polar growth in bacteria. (68, 69) It has also been suggested that even unicellular bacteria grow by intercalation of cell wall material in a diffuse manner. (70) This is based upon an extension of the technique (58, 59) of immunofluorescent labelling. However, it is not at all clear that the substances thus labelled are invariably solid cell wall material. Antigenic capsular slime is present on all or most bacterial surfaces, and may even be secreted as blebs on the wall. (71, 72) This makes it uncertain whether immunofluorescent methods can invariably be trusted to give a clear picture of the cell wall, (42) even apart from the potential hazard, already mentioned above, of treating living bacteria with antibodies against the antigens of their cell envelopes, without very careful control of the effects that this might, by itself, produce upon the sensitive growing-points, by analogy with antibiotics. (48)

THE CELL WALL OF MYXOBACTERIA

The myxobacteria differ from most other bacteria in that they lack the rigid cell wall, and are independent of flagella for motility. (73)

The cell wall cannot be detected in myxobacteria by ordinary staining methods, but

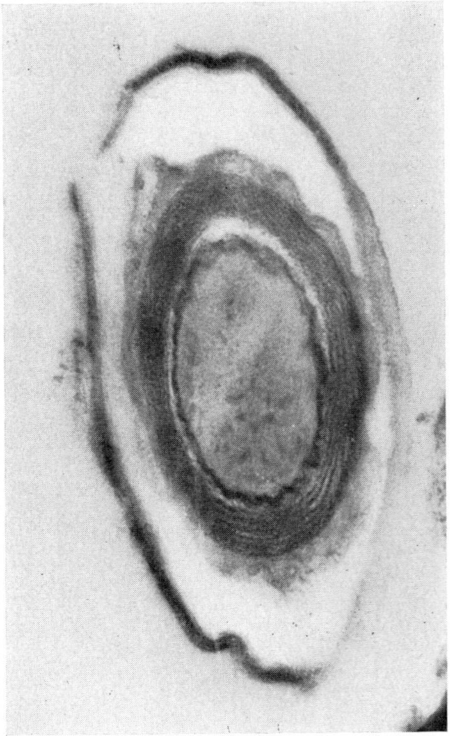

(Reproduced from the Journal of Applied Bacteriology, by permission of Dr C. F. Robinow)

FIG. 25
SPORE MEMBRANES

Electron micrograph of section through the spore of *Bacillus megaterium*, showing many layers of membrane surrounding the cytoplasm and nucleus. ×75,000.

some of its structural characteristics can be inferred from other considerations. [74] Although the cells exhibit a degree of flexibility, their structure is sufficiently rigid to enable them to retain the bacillary form when immersed in fluid. Such strength is unquestionably not possessed by the unprotected cell membrane. Chemical analysis of myxobacterial walls shows them to have a comparable constitution to those of other Gram-negative bacteria, but with a higher proportion of lipid. [75] And a thin, membrane-like wall can be seen in ultra-sections and disrupted cells, under the electron microscope.[75]

Myxobacteria are motile only when in contact with a surface, whether a solid surface or

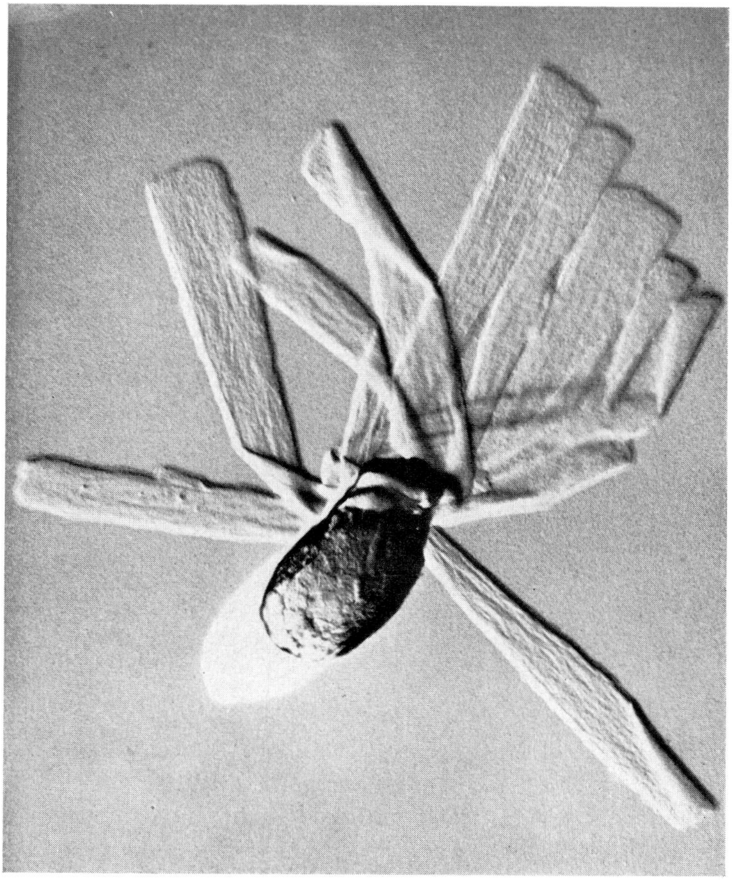

(Reproduced from the Journal of Bacteriology, by permission of Dr L. J. Rode)

FIG. 26
SPORE APPENDAGES

Spore of an anaerobic bacillus with large ribbon-like appendages. It is suggested that these may aid in aerial distribution. ×5,000. See also Figures 92 and 94.

the surface film of a fluid. They show a marked tendency to move along the lines of physical stress in the surface, a phenomenon that is referred to as elasticotaxis. Their mode of progression has been variously described, but appears to the author to be a worm-like action analogous to peristalsis. This implies a muscular activity in the cell wall, which must be capable of contracting circumferentially, to extend the cell, and also longitudinally, to shorten and expand it. Flexion is occasionally shown but probably is not a necessary function of locomotion. [76, 77] Muscular action in so small an organ is not exceptional and is obvious in the locomotory cilia and flagella of many small creatures, including bacteria, but no sign of a fibrous structure can be seen in ultra-sections of the wall.

The microcysts of myxobacteria possess a rigid cell wall more closely resembling that of eubacteria. [78]

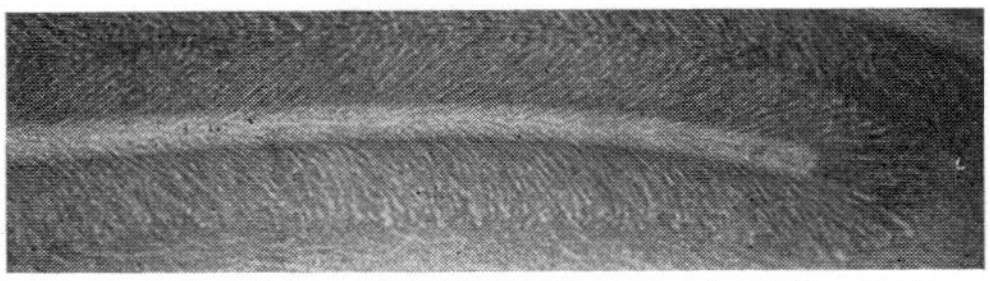

(Reproduced from the Journal of Microbiology, by permission of Dr L. J. Rode)

FIG. 27
SPORE APPENDAGES

The tip of a feather-like appendage, attached to the spore of *Clostridium bifermentans*. ×100,000.

THE SPORE COAT

The resistance of the bacterial endospore to inimical agencies appears to be due mainly to a degree of dehydration, which protects the proteins from denaturation [79, 80, 81] but this peculiar condition must be maintained against the environment by an impenetrable spore coat. The coat is a physically strong structure, although it has a weak point, possibly the germination pore, through which the contents may be ejected when the spore is subjected to acid-hydrolysis; [82] presumably because the proteins then absorb water and become turgid. From the many

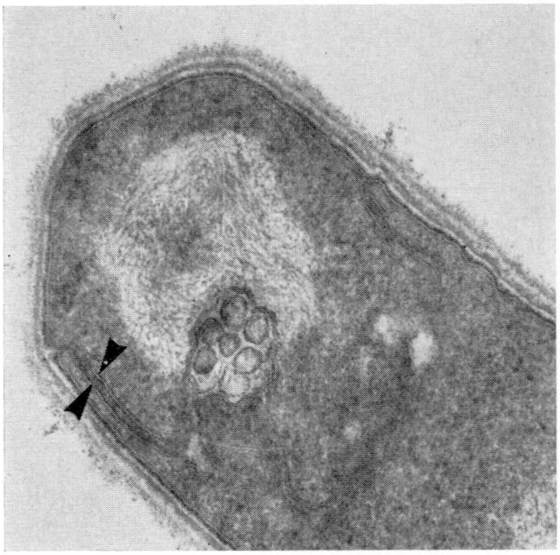

(Reproduced from the Journal of Applied Bacteriology, by permission of Dr P. D. Walker)

FIG. 28
SPORE DEVELOPMENT

Electron micrograph through nucleus (fine strands) and mesosome (rosette) of a developing spore of *Bacillus stearothermophilus*. The arrows show an infolding of the membrane around the spore. ×38,000.

electron micrographs of ultra-sections that have been made, it appears that the spore coat is basically similar to the envelopes of the vegetative bacillus, but the layers are considerably more numerous and convoluted [83, 84] (Figs 17 & 25). Chemically also, it is similar but more complex, [12] having a wider range of amino-acids and, especially, a very high phosphorus content. [85] This is probably correlated with the possession of numerous layers of cell-membrane, in place of the normal single layer, and is one of the obvious, regularly-occurring characters found in the spore coat. The wall of the spore is sometimes further surrounded by a loose exosporium membrane. The exosporium was previously thought to represent the remains of the sporangium or mother cell, but there is evidence to suggest that it is formed between the cell envelopes and the developing spore [86] (Fig. 30). The outer surface of the wall is elaborately sculptured in many species, [87] and may have ribbon-like appendages [88] (Figs 14, 26 & 27). Between the wall or walls and the cytoplasm of the spore lies the thick, multiple cortex, apparently derived from the membrane. [89, 90] It seems reasonable to suppose that such a multiple-folded layer of lipoprotein complex could provide a water-tight seal for the interior of the spore, and that this is a major function of the spore coat. The general structure of the spore envelopes, as shown in ultra-sections, has been confirmed by the freeze-etching technique [91] (Fig. 17).

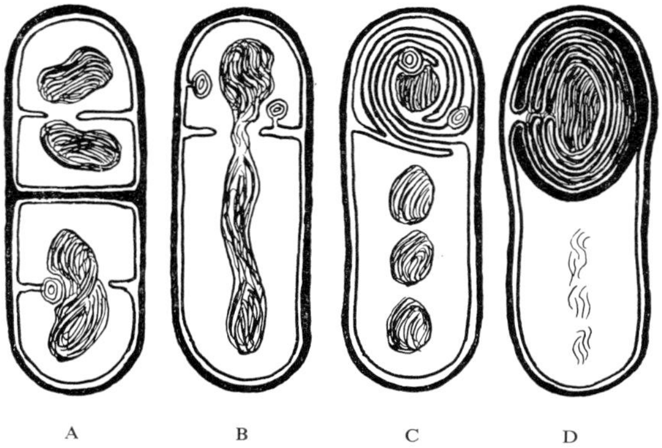

FIG. 29
DIAGRAM OF THE DEVELOPMENT OF THE SPORE MEMBRANES

A, vegetative bacillus, showing a dividing nucleus in the lower cells, associated with a mesosome on the septum.

B, longitudinal fusion nucleus commencing to separate off a presporal area, with mesosomes.

C, the spore nucleus is surrounded by an overfolded membrane, the remainder of the nucleus subdivides.

D, the spore membranes are completed, except for a pore, the excluded nucleoids degenerate (or are discarded). Compare Figures 28, 30, 53, 60, 61, 62.

There have been many studies bearing upon the problem of the spore coat and its components. The work of Fitz-James is especially important. [84, 89, 90, 92, 93] The forespore, from which the mature spore will develop, is enclosed in an envelope that is derived directly from an infolding of the cell membrane (Figs 28 & 29). Several mesosomes are attached to this membrane, and one or more of these may become enclosed in the maturing spore. The thick, membranous cortex is then formed within layers of the forespore envelope, and the wall, on its outer surface (Figs 29 & 30).

CAPSULES

Most bacteria possess a surface layer of mucoid or gelatinous material, which may form a visible capsule or can be as thin as 50 Å, and apparent only in electron-transparent sections. [94, 95]

The phase-contrast studies of Tomcsik [9, 32] have revolutionised the previous concept of bacterial capsules as an amorphous layer of polysaccharide or polypeptide mucus, by demonstrating that they may possess an exceedingly complex structure of alternate striations of polysaccharide and polypeptide, perpendicular to the cell wall; that is parallel to the cross-walls, which are marked by exceptionally massive lamina, as are the poles of the bacilli. The growth of the capsule, like that of the cell wall, occurs mainly at the junctions of the cell wall and cross-walls (Fig. 10).

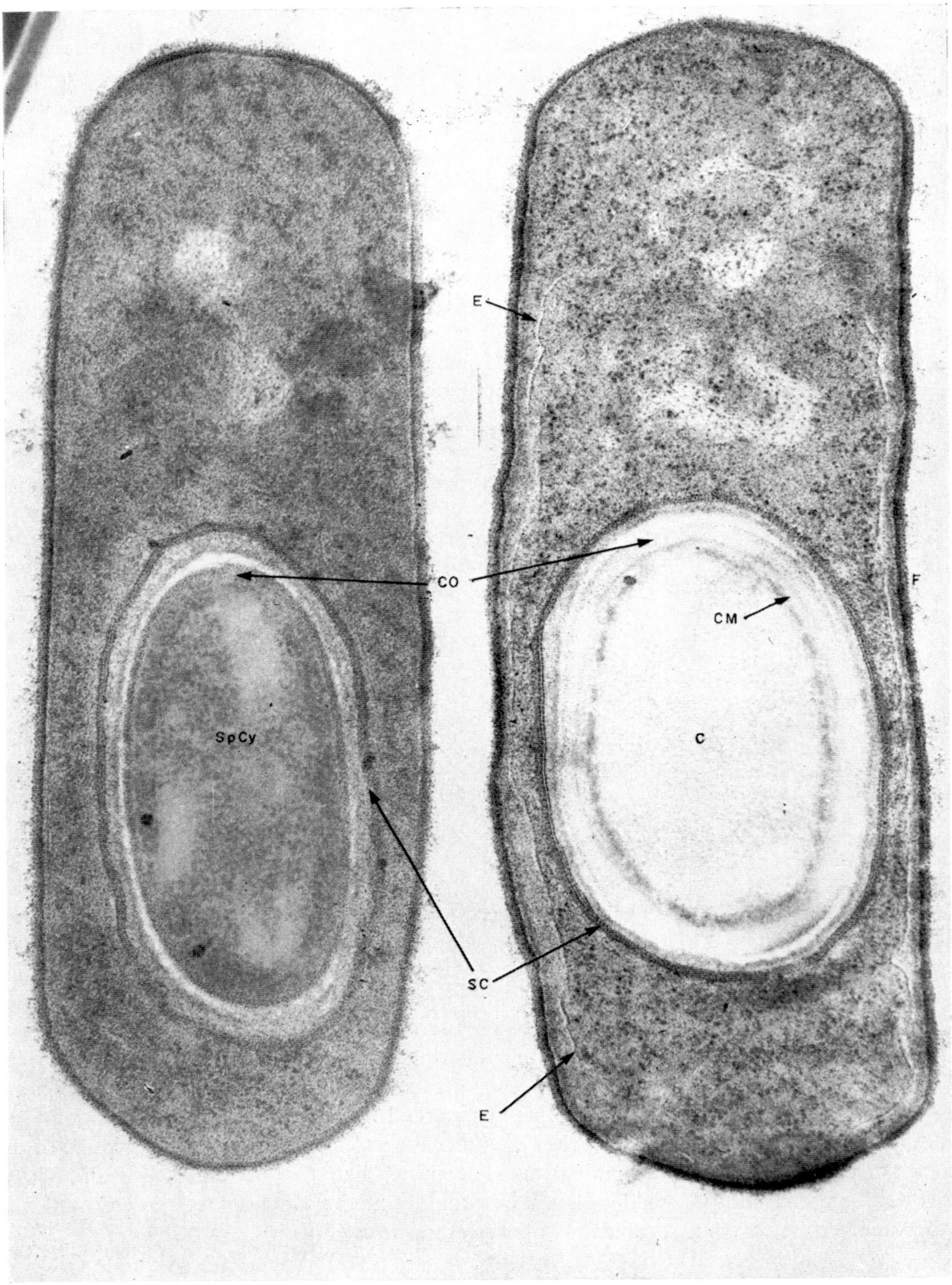

(*Reproduced from the Journal of Applied Bacteriology, by permission of Dr P. D. Walker*

FLAGELLA

Motility in the great majority of bacterial groups is by means of their characteristic flagella (Figs 11, 23, 24 & 34). These may be arranged singly or in small groups at the poles of the cell, in which case they render their possessor very actively motile in a fluid medium, or in larger numbers peritrichously;

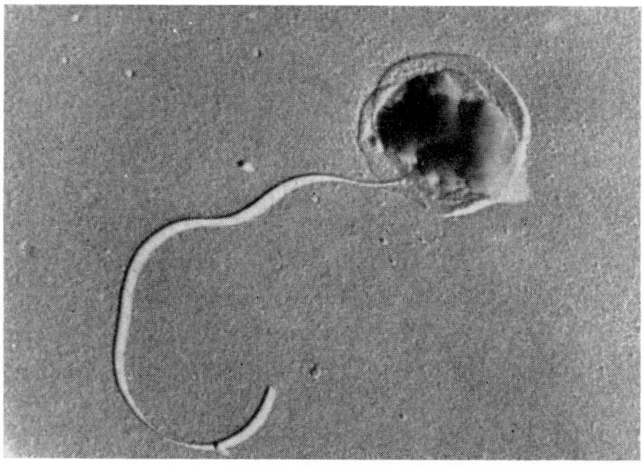

(Electron micrograph by Dr Phyllis Pease)

FIG. 31
FLAGELLUM OF PROTOPLAST

Vibrio. Penicillin-induced L-form protoplast with flagellum. ×15,000.

probably as an adaptation to movement in viscous media or on moist surfaces. Whereas the flagella and cilia of all known cells, except bacteria, have a uniform structure, being composed of nine fibrils arranged around two central fibrils of a different type, the flagella of bacteria consist of a single strand [96] of fibrous protein, flagellin, a member of the keratin-myosin group. [97, 98]

Polar flagella are approximately 30 mμ in diameter in Vibrio and Pseudomonas, but several times as large in Spirillum. Peritrichous flagella may be as small as 12 mμ. [99, 100] Some bacterial flagella are provided with a sheath, which considerably increases their apparent diameter.[101] Because of their small size their mode of action is difficult to determine. They have been described as lashing, but more probably act by waves of contractions passing down their fine coils. The wavelength of these coils varies from 1 to 5 μ and is constant for any bacterial species, but variants in a single strain may have flagella of double the normal wavelength. [102] In addition, each flagellum shows signs of a spiral structure which, in very well defined electron micrographs, gives it the appearance of a rope. [99, 103] This is in accordance with the periodic structure of the fibrous protein of which flagella are composed [97, 98] and so must be a molecular phenomenon.

Flagella point away from the direction of motion, and the rearmost may become twisted together into a spiral thread. Cast-off flagella, in fixed preparations, also tend to knit up into whips in this manner.

The flagella originate in the cell membrane or surface cytoplasm and pass outwards through the cell wall. When the cell wall is removed by digestion with lysozyme, the flagella cease to function but remain attached to the protoplast. [104] This is also

---

FIG. 30
SPORE DEVELOPMENT

Electron micrographs of sections of developing spores of *Clostridium bifermentans*. E, exosporium membrane; CO, cortex; SC, spore coat; CM, cortical membranes; C and SpCy, spore cytoplasm and nuclear structures.

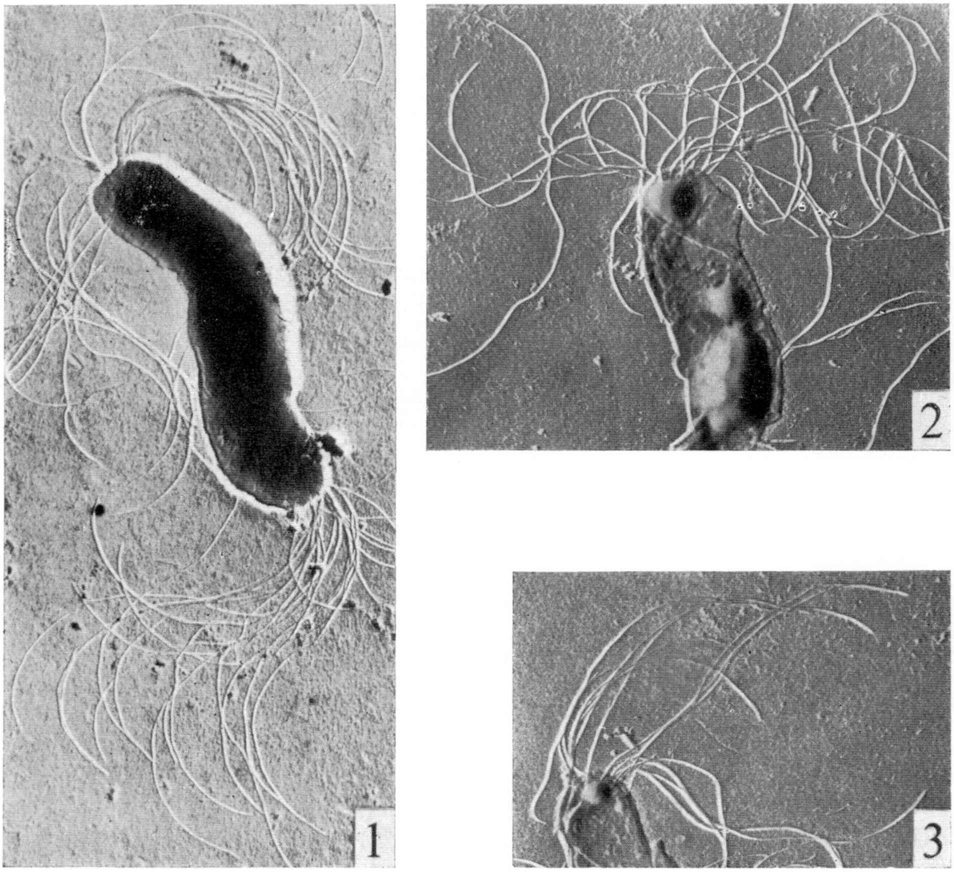

(Reproduced from the Giornale di Microbiologia)

FIG. 32
BLEPHAROPLASTS OF SPIRILLUM

Spirillum serpens. Electron micrographs.
(1) entire spirillum with polar flagella; (2, 3) partially disrupted spirilla showing several flagella attached to each blepharoplast. ×10,000.

true of L-forms produced by the action of penicillin (105) (Fig. 31). Their point of origin is a basal granule of blepharoplast, approximately spherical and rather larger in diameter than the flagellum. Within the granule is a smaller, pointed structure which may be slightly hooked (106, 107) (Figs 32 & 33). In most bacterial genera each flagellum arises from a separate granule, but the flagellar fibrils of spirilla arise in bundles from single granules (108), so that each bundle constitutes a compound flagellum (Figs 32 & 34). This is considered to represent a primitive condition, intermediate between that in

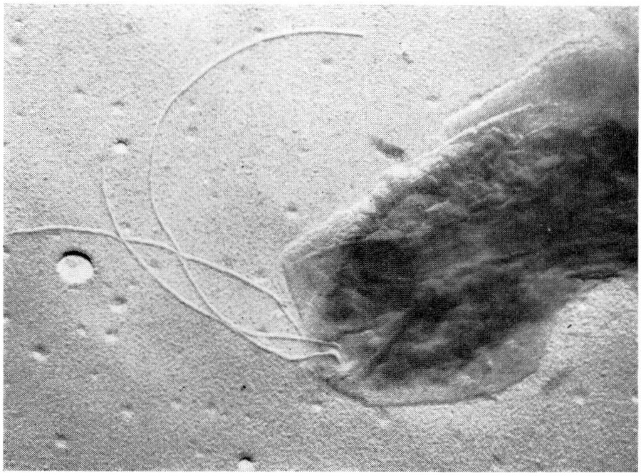

(*Electron micrograph by Dr Phyllis Pease*)

FIG. 33
ATTACHMENT OF FLAGELLA

Spirillum serpens, showing hook-like bases of flagella. ×30,000.

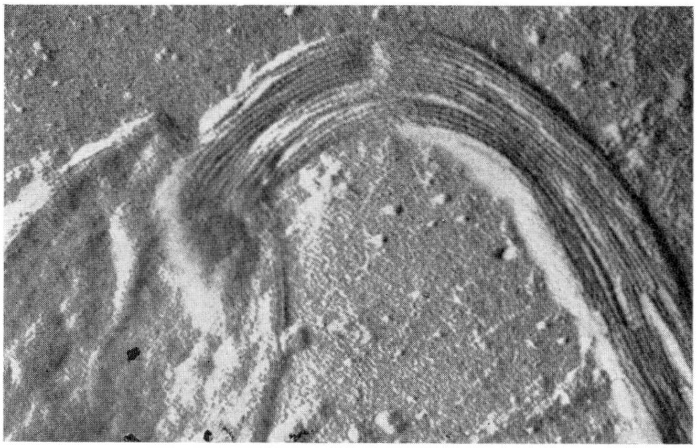

(*Electron micrograph by Dr Phyllis Pease*)

FIG. 34
COMPLEX FLAGELLUM

Complex flagellum of *Spirillum* sp. ×40,000.

typical bacteria and in the flagellate protista (Chap. 9).

The microcysts and spores of bacteria are devoid of flagella. These commence to grow at germination, usually at the pole remote from the growing-point of the cell. [(63)] There is a great deal of evidence to suggest that flagella in dividing bacteria are not distributed equally between the two daughter cells, but that a majority remain attached to one of these [(2, 62, 64, 66, 67, 109)] (Figs 22, 23 & 24).

The possession of flagella is the most important single factor indicating relationship between bacteria and the flagellate protista, rather than with the blue-green algae, as is often suggested, and much of the argument concerning evolutionary relationships between bacteria is based upon the evidence of these structures (Chap. 9).

In addition to flagella, some bacteria possess fimbriae. [(99, 110)] These superficially resemble flagella, but are shorter and straight. They may have an invasive function in some pathogenic Salmonellas [(111)] but are not well understood.

The axial filament of spirochaetes may be regarded as a peculiar type of sub-polar flagellum. [(12)] Arising near the tip, it is arranged along the axis of the coiled body of the spirochaete (Fig. 35). The action of the fibrous strand presumably throws the spiral body into a series of contractions, and thus propels the spirochaete by a motion similar to that of the normal flagellum, except that the roles of body and flagellum are reversed.

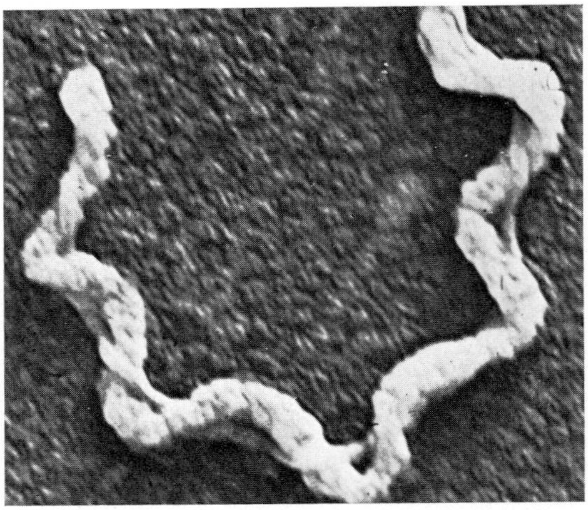

(Reproduced from the Journal of Pathology and Bacteriology, by permission of Dr J. W. Czekalowski)

FIG. 35
AXIAL FILAMENT

Electron micrograph of a leptospira, showing the axial filament (a modified flagellum) attached near one pole. ×83,000.

SUMMARY

Eubacteria have a strong, rigid cell wall of at least two and sometimes three layers, 100 Å thick; within this lies the semi-permeable membrane. The composition of the wall is not identical in all species, but the external layers, which may be elaborately sculptured, contain lipid, protein and polysaccharide macromolecular complexes. The rigid central layers have an important mucopolymer component. The cell membrane is a lipid-protein complex, as in other cells, and has attached to it the mesosomes, folded invaginations that are apparently concerned with energy transfer in the division of both cell and nucleus. The wall is secreted by the membrane in certain well-marked areas, which correspond to the poles of cells and the points of division. As bacteria may have up to 20 or more cells in each bacillus, they can have numerous growing-points. Unicellular ('smooth') Gram-negative bacteria frequently have only one growing-point, at one pole, so that all or most of the flagella pass to one daughter cell.

The cell wall of myxobacteria is flexible, but is chemically similar to that of eubacteria.

Capsules are found in almost all bacteria, but vary greatly in thickness. They are less simple than they appear, and have been shown, in some cases, to consist of laminations of polysaccharide and polypeptide, at right-angles to the cell wall.

The cell envelopes have a strong basophilic element, especially in the vicinity of the growing points with their associated mesosomes; these whole areas appear as stainable granules by many simple techniques, and have been a cause of confusion, because of their reaction with diagnostic stains (*e.g.* for mitochondria).

The cell envelopes of spores are similar to those of vegetative cells, but are more complex. The wall is often double, and the lipid-containing membrane, reflected many times around the cytoplasm, provides a watertight seal for the partially dehydrated interior.

The flagella arise in or near the membrane, from a hooked base, embedded in a blepharoplast granule. Each flagellum is a single strand of flagellin, a fibrous protein, and is in the form of a loose spiral, down which pass waves of contraction. Bacterial flagella point away from the direction of movement.

Flagella of spirilla are thicker than those of other bacteria, and are sometimes composed of several strands beating together. The axial filaments of spirochaetes appear to be a specialised type of flagellum.

Some bacteria possess fimbriae that resemble flagella, but are not organs of motility.

REFERENCES

1. Pijper, A. (1946). *J. Path. Bact.* **58**, 325.
2. Bisset, K. A. (1951). *J. gen. Microbiol.* **5**, 155.
3. Cole, R. M. (1965). *Bact. Rev.* **29**, 326.
4. Bisset, K. A. (1953). Symposium citologia batterica, *6th Int. Congr. Microbiol.* 9.
5. Bisset, K. A. (1957). *Ergebn. Mikrobiol. ImmunForsch. exp. Ther.* **30**, 1.
6. Bisset, K. A. & Hale, C. M. F. (1953). *Expl Cell Res.* **5**, 449.
7. Tuffery, A. A. (1954). *J. gen. Microbiol.* **10**, 342.
8. Bisset, K. A. (1954). *J. Bact.* **67**, 41.
9. Tomcsik, J. & Guex-Holzer, S. (1954). *J. gen. Microbiol.* **10**, 97.
10. Holdsworth, E. S. (1952). *Biochim. Biophys. Acta,* **8**, 110.
11. Holdsworth, E. S. (1952). *Biochim. Biophys. Acta,* **9**, 19.
12. Salton, M. R. J. (1964). *The Bacterial Cell Wall.* New York: Elsevier.
13. Cummins, C. S. & Harris, H. (1956). *J. gen. Microbiol.* **14**, 583.
14. Bisset, K. A. & Vickerstaff, J. (1967). *Nature, Lond.* **215**, 1286.
15. Houwink, A. L. (1953). *Biochim. Biophys. Acta,* **10**, 36.
16. Nermut, M. V. & Murray, R. G. E. (1967). *J. Bact.* **93**, 1949.
17. de Boer, W. E. & Spit, B. J. (1964). *Antonie van Leeuwenhoek*, **30**, 239.
18. Giesbrecht, P. & Drews, G. (1966). *Arch. Mikrobiol.*, **54**, 297.
19. Glauert, A. M. (1962). *Br. med. Bull.* **18**, 245.
20. Steed, P. & Murray, R. G. E. (1965). *Can. J. Microbiol.* **12**, 263.
21. Walker, P. D. & Baillie, A. (1968). *J. appl. Bact.* **31**, 108.
22. Chapman, G. B. & Hillier, J. (1953). *J. Bact.* **66**, 362.
23. Remsen, C. & Lundgren, D. G. (1966). *J. Bact.* **92**, 1765.
24. Nermut, M. V. (1967). *J. gen. Microbiol.* **49**. 503.

25. MURRAY, R. G. E., STEED, P. & ELSON, H. E. (1965). *Can. J. Microbiol.* **11**, 547.
26. BISSET, K. A. (1966). *G. Microbiol.* **14**, 5.
27. ROGERS, H. J. & PERKINS, H. R. (1968). *Bacterial Cell Walls & Membranes.* London: Spon.
28. ROBINOW, C. F. & MURRAY, R. G. E. (1953). *Expl Cell Res.* **4**, 390.
29. NEWTON, B. A. (1955). *J. gen. Microbiol.* **12**, 226.
30. WEIBULL, C. (1953). *J. Bact.* **66**, 688.
31. WEIBULL, C. (1953). *J. Bact.* **66**, 696.
32. TOMCSIK, J. & GEUX-HOLZER, S. (1954). *J. gen. Microbiol.* **10**, 317.
33. BISSET, K. A. (1948). *J. gen. Microbiol.* **2**, 83.
34. PIJPER, A. (1938). *J. Path. Bact.* **47**, 1.
35. HENRY, H. & STACEY, M. (1943). *Nature, Lond.* **151**, 671.
36. HENRY, H., STACEY, M. & TEECE, E. G. (1945), *Nature, Lond.* **156**, 720.
37. MITCHELL, P. & MOYLE, J. (1950). *Nature, Lond.* **166**, 218.
38. HOFFMAN, H. (1951). *Nature, Lond.* **168**, 464.
39. GUTSTEIN, M. (1924). *Zentbl. Bakt. ParasitKde, I, Orig.* **93**, 393.
40. HALE, C. M. F. (1953). *Lab. Pract.* **2**, 115.
41. FITZ-JAMES, P. C. (1966). *Symp. Soc. gen. Microbiol.* **15**, 369.
42. ROGERS, H. J. (1966). *Symp. Soc. gen. Microbiol.* **15**, 186.
43. GOULA, E. A., BUTLER, T. F., KING, R. D. & SMITH, G. L. (1967). *Can. J. Microbiol.* **13**, 1499.
44. BISSET, K. A. & HALE, C. M. F. (1951). *J. gen. Microbiol.* **5**, 592.
45. GOULA, E. A. & SMITH, G. L. (1965). *J. Bact.* **90**, 1054.
46. BISSET, K. A. (1938). *J. Path. Bact.* **47**, 223.
47. BERGERSEN, F. J. (1953). *J. gen. Microbiol.* **9**, 26.
48. BERGERSEN, F. J. (1953). *J. gen. Microbiol.* **9**, 353.
49. GLAUERT, A. M., BRIEGER, E. M. & ALLEN, J. M. (1961). *Expl. Cell Res.* **22**, 73.
50. COHEN-BAZIRE, G., KUNISAWA, R. & POINDEXTER, J. S. (1966). *J. gen. Microbiol.* **42**, 301.
51. VANDERWINKEL, E. & MURRAY, R. G. E. (1962). *J. Ultrastruct. Res.* **7**, 185.
52. VAN ITERSON, W. & LEENE, W. (1964). *J. Cell Biol.* **20**, 361.
53. VAN ITERSON, W. (1961). *J. Biophys. Biochem. Cytol.* **9**, 183.
54. ELLAR, D. J., LUNDGREN, D. G. & SLEPECKY, R. A. (1967). *J. Bact.* **94**, 1189.
55. RYTER, A. & JACOB, F. (1964). *Annls Inst. Pasteur, Paris,* **107**, 384.
56. BEATON, C. D. (1968). *J. gen. Microbiol.* **50**, 37.
57. GLAUERT, A. M. & HOPWOOD, D. A. (1959). *J. Biophys. Biochem. Cytol.* **6**, 515.
58. COLE, R. M. & HAHN, J. J. (1962). *Science, N.Y.* **135**, 722.
59. VAN ITERSON, W. & LEENE, W. (1964). *J. Cell Biol.* **20**, 377.
60. WALKER, P. D. & BAILLIE, A. (1968). *J. appl. Bact.* **31**, 108.
61. PENNINGTON, D. (1950). *J. Bact.* **59**, 617.
62. HALE, C. M. F. & BISSET, K. A. (1958). *J. gen. Microbiol.* **18**, 688.
63. BISSET, K. A. & HALE, C. M. F. (1951). *J. gen. Microbiol.* **5**, 150.
64. BISSET, K. A. & PEASE, P. E. (1957). *J. gen. Microbiol.* **16**, 382.
65. QUADLING, C. & STOCKER, B. A. D. (1962). *J. gen. Microbiol.* **28**, 257.
66. QUESNEL, L. B. (1966). *Nature, Lond.* **211**, 659.
67. WILSON, C. E., DONATI, E. J., PETRALI, J. P., VUICICH, J. V. & STERNBERGER, L. A. (1966). *Expl molec. Path.* Suppl. **3**, 44.
68. MALEK, I., VOSKYOVA, L., WOLF, A. & FIALA, J. (1954). *Čslká Biol.* **3**, 135.
69. ADLER, H. I. & HARDIGREE, A. A. (1964). *J. Bact.* **87**, 1240.
70. BEACHEY, E. H. & COLE, R. M. (1966). *J. Bact.*, **92**, 1245.
71. KNOX, K. W., VESK, M. & WORK, E. (1966). *J. Bact.* **92**, 1206.
72. BAYER, M. E. (1967). *J. Bact.* **93**, 1104.
73. KRZEMIENIEWSKI, H. & S. (1928). *Acta Soc. Bot. Pol.* **5**, 46.
74. KLIENEBERGER-NOBEL, E. (1947). *J. gen. Microbiol.* **1**, 22.
75. MASON, D. J. & POWELSON, D. (1958). *Biochim. Biophys. Acta*, **29**, 1.
76. STANIER, R. Y. (1942). *Bact. Rev.* **6**, 143.
77. STANIER, R. Y. (1942). *J. Bact.* **44**, 405.
78. LOEBECK, M. E. & ORDAL, E. J. (1957). *J. gen. Microbiol.* **16**, 76.

79. BISSET, K. A. (1950). *Nature, Lond.* **166**, 431.
80. ROSS, K. F. A. & BILLING, E. (1957). *J. gen. Microbiol.* **16**, 418.
81. MURRELL, W. G. & SCOTT, W. J. (1957). *Nature, Lond.* **179**, 481.
82. BISSET, K. A. & HALE, C. M. F. (1951). *J. Hyg. Camb.* **49**, 201.
83. MAYALL, B. H. & ROB, C. (1957). *J. appl. Bact.* **20**, 333.
84. TOKUYASU, K. & EICHI, Y. (1959). *J. Biophys. Biochem. Cytol.* **5**, 129.
85. FITZ-JAMES, P. C. (1955). *Can. J. Microbiol.* **1**, 502.
86. WALKER, P. D., THOMASON, R. O. & BAILLIE, A. (1967). *J. appl. Bact.* **30**, 444.
87. BRADLEY, D. E. & WILLIAMS, D. J. (1957). *J. gen. Microbiol.* **17**, 75.
88. RODE, L. J., CRAWFORD, M. A. & WILLIAMS, M. G. (1966). *J. Bact.* **93**, 1160.
89. YOUNG, E. & FITZ-JAMES, P. C. (1962). *J. Cell Biol.* **12**, 115.
90. FITZ-JAMES, P. C. (1962). *J. Bact.* **84**, 104.
91. REMSEN, C. C. (1966). *Arch. Mikrobiol.* **54**, 266.
92. FITZ-JAMES, P. C. (1960). *J. Biophys. Biochem. Cytol.* **8**, 507.
93. ELLAR, D. J. & LUNDGREN, D. G. (1966). *J. Bact.* **92**, 1748.
94. WILKINSON, J. F. (1958). *Bact. Rev.* **22**, 46.
95. GLAUERT, A. M. (1962). *Br. med. Bull.* **18**, 245.
96. RHODES, M. E. (1965). *J. Bact.* **29**, 442.
97. ASTBURY, W. I., BEIGHTON, E. & WEIBULL, C. (1955). *Symp. Soc. exp. Biol.* **9**, 282.
98. KERRIDGE, D. (1961). *Symp. Soc. gen. Microbiol.* **11**, 41.
99. HOUWINK, A. L. & VAN ITERSON, W. (1950). *Biochim. Biophys. Acta*, **5**, 10.
100. WEIBULL, C. (1950). *Acta chem. scand.* **4**, 268.
101. VAN ITERSON, W. (1953). Symposium citologia batterica, *6th Int. Congr. Microbiol.* 24.
102. PIJPER, A., NESER, M. & ABRAHAM, G. (1956). *J. gen. Microbiol.* **14**, 371.
103. LABAW, L. W. & MOSLEY, V. M. (1954). *Biochim. Biophys. Acta*, **15**, 325.
104. WEIBULL, C. (1953). *J. Bact.* **66**, 688.
105. PEASE, P. E. (1957). *G. Microbiol.* **3**, 44.
106. GRACE, J. B. (1954). *J. gen. Microbiol.* **10**, 325.
107. ABRAM, D., VATTER, A. E. & KOFFLER, H. (1966). *J. Bact.* **91**, 2045.
108. BISSET, K. A. (1960). *G. Microbiol.* **8**, 193.
109. BISSET, K. A. & HALE, C. M. F. (1960). *J. gen. Microbiol.* **22**, 536.
110. DUGUID, J. P., SMITH, I. W., DEMPSTER, G. & EDMUNDS, P. N. (1955). *J. Path. Bact.* **70**, 335.
111. DUGUID, J. P. & GILLIES, R. R. (1957). *J. Path. Bact.* **74**, 397.
112. CZEKALOWSKI, J. W. & EAVES, G. (1955). *J. Path. Bact.* **69**, 129.

CHAPTER 4

The Bacterial Nucleus

HISTORICAL

The existence of the bacterial nucleus was long denied, mainly upon the evidence that it is not readily demonstrated in preparations fixed and stained according to standard bacteriological procedures. Good descriptions of the nuclear apparatus, so far as it can be revealed by the light microscope, were published from time to time, but were ignored by almost all bacteriologists until quite recently.

It has already been mentioned, in the Introduction, that the observations upon eubacteria, made in the nineteen-forties and fifties, were preceded by a comparable demonstration of the nucleus in myxobacteria. These micro-organisms received the cytological treatment usually denied to eubacteria, and their nuclear material can be revealed by simple staining techniques, which that of eubacteria often cannot. Since they are no less small than eubacteria, this fact disposed of the theory that bacteria are not large enough to warrant possession of a nucleus.

A revolution in bacterial cytology resulted from the adoption of the technique of acid-hydrolysis, as a preliminary to staining. This was originally applied in the process of the Feulgen reaction for nucleic acids, which was used successfully by Stille and by Piekarski [1, 2] in 1937, and by many others at about the same date, to demonstrate nuclear structures in bacteria, although a variant had been used by Gutstein [3] in 1924, without attracting much attention at the time. Piekarski also discovered that after acid-hydrolysis the nucleus stained clearly with Giemsa (Fig. 4). This technique was adopted by Robinow, [4] whose work attracted considerable attention, and was, in fact, the first description of typical, paired bacterial nucleoids to obtain general credence, although by no means the first to be published. Some simple, but accurate figures appeared in a description of an insect pathogen by Paillot [5] as early as 1919, and other examples might be quoted.

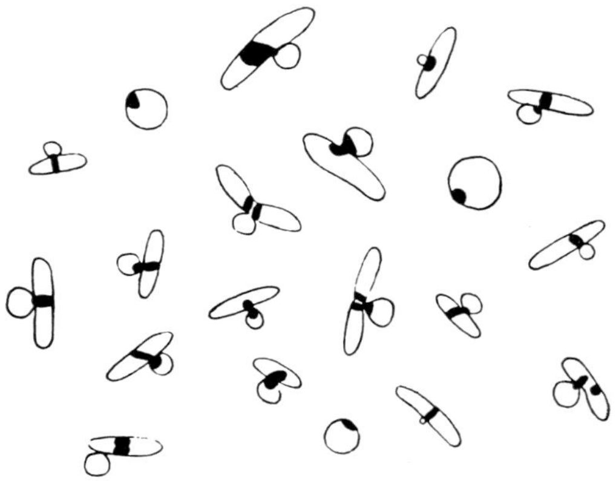

(After Stoughton. Reproduced from the Proceedings of the Royal Society)

FIG. 36
MATURATION OF THE RESTING CELL IN *BACTERIUM MALVACEARUM*

The microcyst is formed from the secondary nuclear phase, and is extruded laterally from the mother cell.

A general understanding of the nature of the

vegetative nucleus in the commoner eubacteria became widespread after Robinow's paper in 1942, although the technical difficulties of demonstrating it in practice have always been a cause of disappointment and misunderstanding.

Claims have been made, from time to time, that a classical mitosis exists in bacteria. These have never stood up to investigation, and have commonly been based upon a misunderstanding of the complex structure that is found in the cell envelopes, even of quite small bacteria (Chap. 3).

The observation that the bacterial nucleus undergoes a definite cycle, in the process of maturation of the resting stage, almost antedates the clarification of the nature of the vegetative nucleus. In the early thirties Badian [6, 7, 8] gave an account of this, as it concerned both the endospore of bacilli and the microcyst of myxobacteria; Krzemieniewska [9] illustrated very similar processes in cytophagas, and a study by Stoughton [10, 11] on a plant-pathogenic bacterium, showed that this was not confined to bacteria with well-recognised spores and cysts (Fig. 36). In the decade following Robinow's demonstration of the vegetative nucleus, the maturation of the resting nucleus was further studied in myxobacteria and sporing bacilli by Klieneberger-Nobel, [12, 13] Bisset, [14] Grace [15] Flewett [16] and others, and an almost identical mechanism was shown to exist in the Actinomycetales by Klieneberger-Nobel [17] and Morris, [18, 19] and in non-sporing eubacteria by Bisset [20] (Chap. 5). Bisset also claimed that the resting nucleus was vesicular in character, compared with the vegetative form, and suggested that the maturation embodied a fusion and reduction process. [20, 21, 22] This reduction process was clearly confirmed by Pulvertaft, [23] using phase-contrast microscopy on living bacteria, and shortly afterwards in 1956, Mason and Powelson [24] made the first convincing photomicrographs of living and dividing vegetative nuclei, by the same method. Their pictures were very similar indeed to those obtainable by the acid-Giemsa technique, and this may be regarded as having closed the chapter of classical studies on the bacterial nucleus by light-microscopy. The future lay with the electron microscope and the ultra-microtome (Fig. 37).

After a rather long delay, caused by an unexpected difficulty in finding suitable fixation methods for embedding, the ultrastructure of the nucleus was clearly demonstrated [25, 26, 27] by several workers, in the early nineteen-sixties. It appeared as a bundle of fibrils, folded over upon itself, and was soon afterwards shown to be a single continuous molecular strand, about 1,100 to 1,400 μ in length, by Cairns [28, 29] who employed, not electron microscopy but an ingenious autoradiographic technique (Figs 27, 28 & 38).

Perhaps it is not out of place, at this juncture, to point out that the problem of the morphology and division of the bacterial nucleus is by no means solved. There is too big a gap between the information obtainable from stained preparations or phase-contrast, on one hand, and ultra-sections on the other; they cannot be completely correlated, and few attempts to do so have, in fact, been made. Actually, the schemes of Fitz-James [30] and especially of Tulasne and Collette Vendrély [31] based on visible-light appearances and published in 1954, before the newer evidence was available, accord with it very well (Fig. 46). Kellenberger's later synthesis [25] was an avowed attempt to explain the appearances seen in sections and does little to elucidate the visible-light figures. However, these

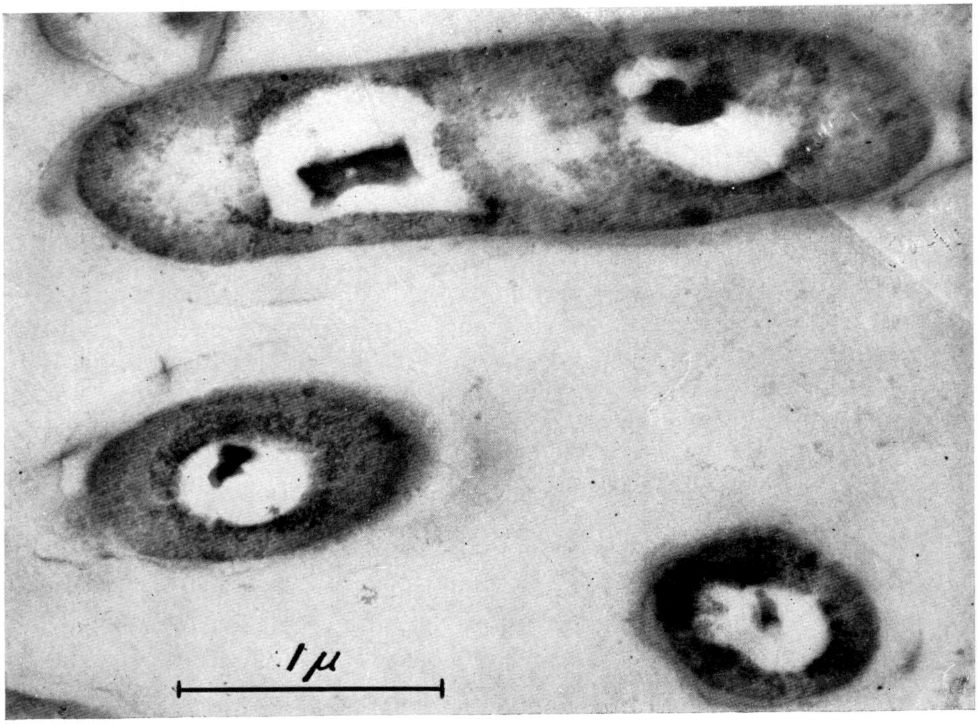

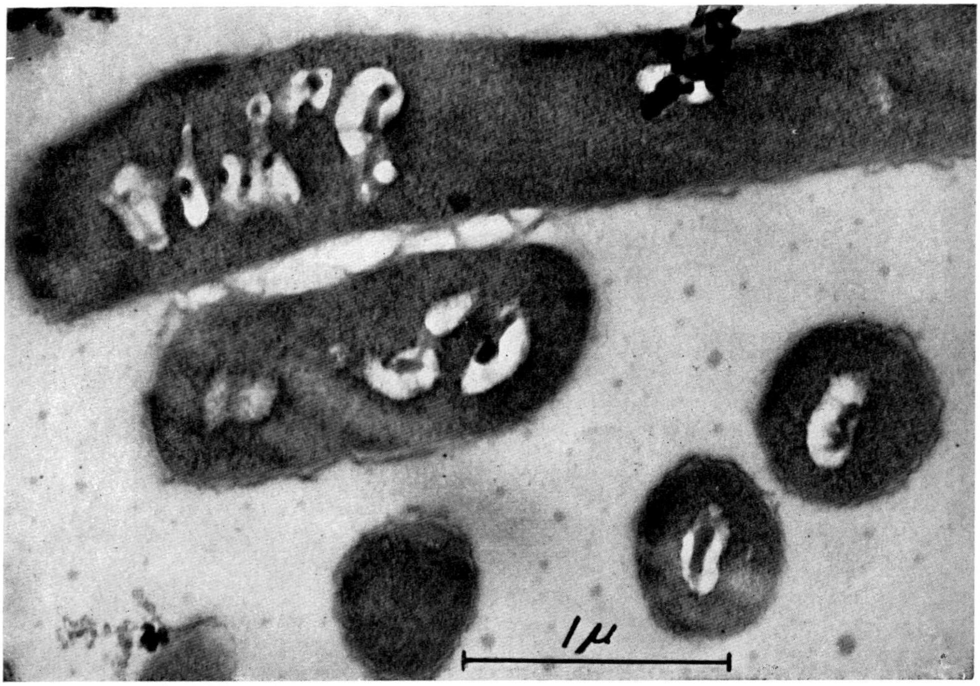

(*Reproduced from Biochimica et Biophysica Acta by permission of Drs Birch-Andersen Maaloe and S'östrand*)

various concepts are not mutually exclusive, as will be apparent when they are discussed more fully (pp. 63, 64).

THE VESICULAR NUCLEUS

The bacterial nucleus, like those of other types of cell, may appear in a variety of different guises. It is probably even more protean than most, but the changes of form which it undergoes are paralleled by similar processes that have been observed in algae and fungi, or even in more complex creatures [6-22] (Fig. 39).

The form of nucleus usually regarded as the standard equipment of a cell, a roughly spherical, vesicular structure, is found in most bacteria at some stage of their life-history. It is best known in the endospore but the spore nucleus is not, in fact, unique or even exceptional as a bacterial resting nucleus. Similar structures were early recognised in myxobacterial microcysts, and later in Actinomycetales. This form of nucleus was not the first to be described in bacteria, nor is it the easiest to demonstrate. Frequently it occurs in resting conditions of the cell, when metabolic activity is low, and the nucleic acid content, upon which nuclear staining reactions depend, is considerably reduced. Hence, in the Bacteriaceae, its presence was undetected for some years after the appearance of the active nucleus, in this type of bacterium, was well recognised (Figs 4 & 40).

In those bacteria which possess spherical nuclei in the active condition it is more readily demonstrable. It is found in the active

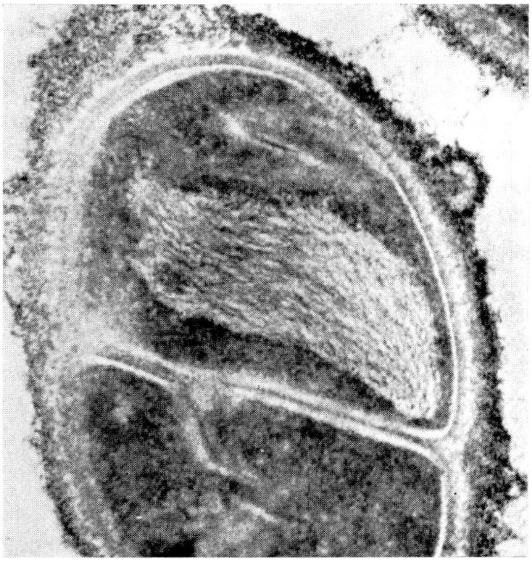

(Reproduced from the Journal of Biophysical and Biochemical Cytology, by permission of Drs W. van Iterson and C. F. Robinow)

FIG. 38
BACTERIAL NUCLEUS

Electron micrograph of a section of one cell in a septate coccus. The nucleus is clearly seen as a slightly twisted hank of DNA thread. ×54,000.

FIG. 37
SECTIONS OF BACTERIAL NUCLEI

Electron micrographs of ultra-thin sections of *E. coli*. The material has probably suffered some distortion in the process of fixation in osmium tetroxide solution and embedding in synthetic resin, but nevertheless shows the absence of cross-walls in this type of bacterium, and the nuclear bodies in the form of short rods in both longitudinal and transverse sections. Exactly as in stained preparations, the nuclei appear singly or in pairs towards each end of the cells. The complex structures in the lower micrograph may represent a coiled chromosome.

form in some, although by no means all cocci, in the small cells which comprise the bacillary forms of corynebacteria and mycobacteria, and in Azotobacter. [32-36] In the small bacteria it appears spherical and homogeneous in stained preparations, but in Azotobacter, which is considerably larger, it can be seen to possess a vesicular structure, consisting of an unstained vacuole surrounded by chromatinic granules. It may be supposed that the same structure would be found in the nuclei of the smaller cells, were it possible to resolve them with the microscope. There is evidence that not all these granules are cytochemically identical [35] (Fig. 6).

The nuclei of mycobacteria were described as Feulgen-positive granules, regularly arranged along the length of the bacillus, before it was realised that the bacillus is multicellular, and that each granule was a cell nucleus. [34] Some confusion has also resulted from identification of these granules with those which appear in the well-known granular or beaded effect seen in heat-fixed preparations of *M. tuberculosis*. The latter are, in fact, merely the shrunken cell contents.

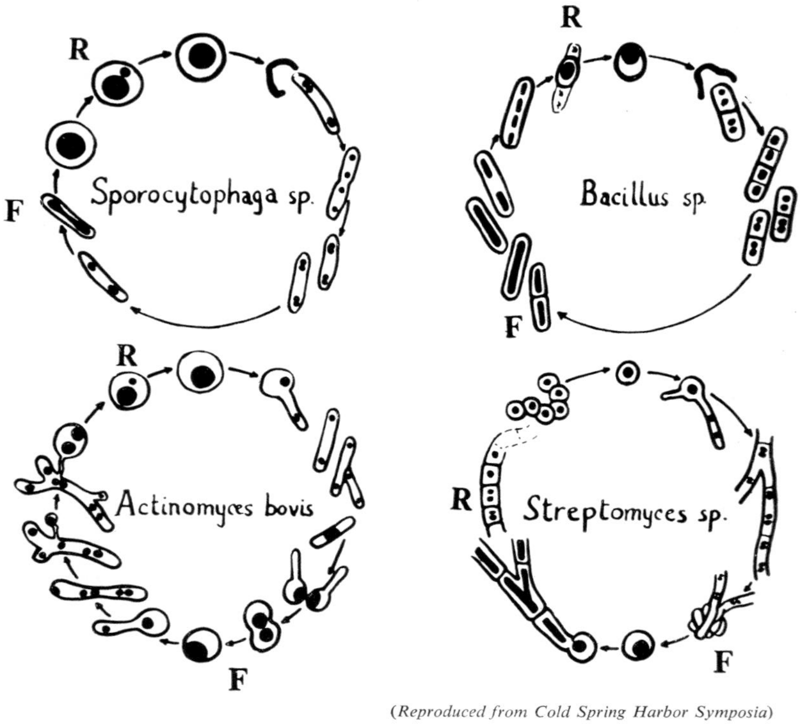

(*Reproduced from Cold Spring Harbor Symposia*)

FIG. 39

NUCLEAR CYCLES IN A VARIETY OF BACTERIA

F-fusion and R-reduction, in each case. In *Sporocytophaga* and in non-sporing eubacteria there is no diploid generation. In *Bacillus* there is a brief diploid (or polyploid) generation. In the higher forms, a secondary, diploid mycelium.

The metachromatic granules of *Corynebacterium diphtheriae* have also been identified with nuclei by some workers, and disproved by others. They are not seen except in dried preparations, and are artefacts consisting of an aggregate of stainable material within the larger, terminal cells of the bacillus [34] (Figs 1 & 41).

Where the spherical nucleus is found only in the resting stages of the bacterium it is often more obviously vesicular and stains eccentrically. The main body of the nucleus stains poorly or not at all, and may be difficult to distinguish from the cytoplasm of the cell (Fig. 40). The stainable portion of the resting nucleus is sometimes merely a crescentic portion of the outline, but may take the form of a single or double spherical body at one edge. Often this is the only portion of the nucleus that can be resolved, and the appearance is of a small, double, spherical body lying eccentrically in the cell. In the case of myxobacteria and cytophagas the stainable portion may be tadpole-shaped. Resting nuclei have been observed in electron micrographs.[37]

The spore nucleus is a vesicular structure, like other resting nuclei. Its component materials appear to be in a condition of partial dehydration, and it is by this means that its resistance to denaturation by heat or chemical agents is achieved. [38, 39] The effect of acid-hydrolysis is to weaken the spore-coat and permit partial rehydration, so that the nuclear material swells, and is forced from its natural position to lodge at the periphery of the cell (Figs 42 & 43). The appearances resulting from this reaction have been variously described as 'extracytoplasmic', 'peripheral' or 'crescentic' nuclei, but they are now generally agreed to be artefacts. [40, 43] Spores of different species of bacteria react in various degrees to this treatment, and produce diverse appearances.

The identity of the spherical nucleus that is found in active cultures of some genera, with that which is confined to the resting stages of most types of bacteria, is not certain. Bacteria, whose active nucleus is spherical in appearance but small, may possess a resting nucleus that is larger and more obviously vesicular (Chap. 6). What appears to be a nucleus of this type was subjected to a very careful examination by van Iterson and Robinow, [26] who compared the appearances seen in stained and sectioned preparations. There seems to be little doubt that, fundamentally, it was exactly similar to other bacterial nuclei, and consisted of a much-folded, simple thread. Quite probably, the spherical appearance implies nothing more than inadequate resolution.

In the cells of most classes of living organism the nucleus returns from the chromosomal to the resting condition between each division, but in bacteria an active condition may be retained throughout the period of vegetative reproduction, and the resting nucleus is restored only when reproduction ceases. Because the active condition of the nucleus is so much more readily demonstrable it was formerly supposed that the nuclear material preserved an organised form only in young cultures, and became disintegrated and distributed throughout the cytoplasm when cultures were more than a few hours old. This, however, is a fallacy.

The resting cells of most bacteria have been so little studied that bacteriologists tend to be unfamiliar with their appearance, and are often unaware that their morphology may be much more distinctive than that of the better-known vegetative stages. Species of Bacteriaceae that are supposedly identical in

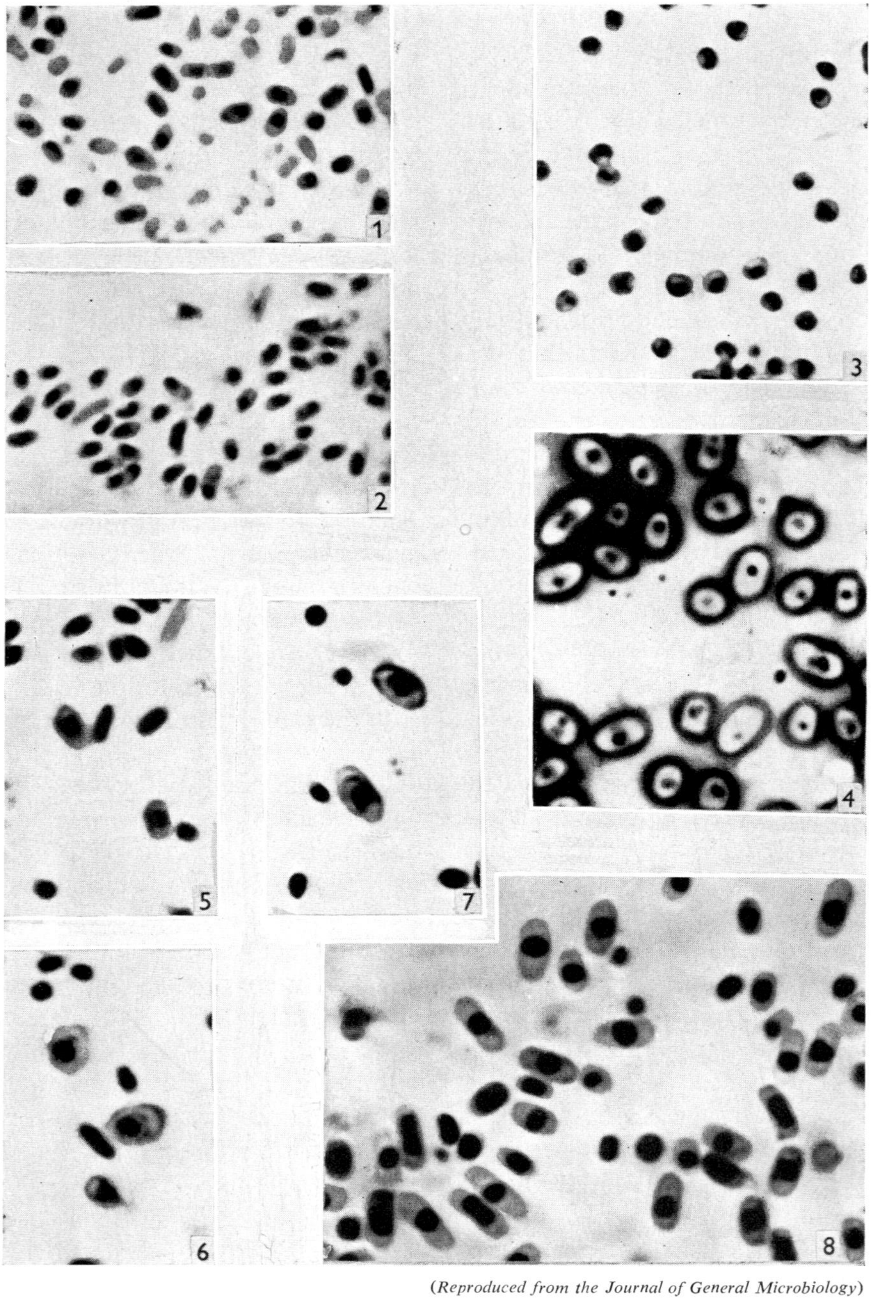

(Reproduced from the Journal of General Microbiology)

FIG. 40

MICROCYSTS OF *BACTERIACEAE*

The appearance of the resting cell and resting nucleus may be very distinctive, even in bacteria of which the vegetative stages are similar. Acid-Giemsa, ×3000.

(1, 2) *E. coli* type. Small, oval cells with an eccentrically staining nucleus. *Proteus* and most *Salmonella* are similar.

(3, 4) *Aerobacter* type, much larger, with a small, central nucleus. ((4) is stained by methylene-blue-eosin.)

(5-7) The large microcysts of *S. typhi*.

(8) *Shigella schmitzii*, large oval cells with a central nucleus.

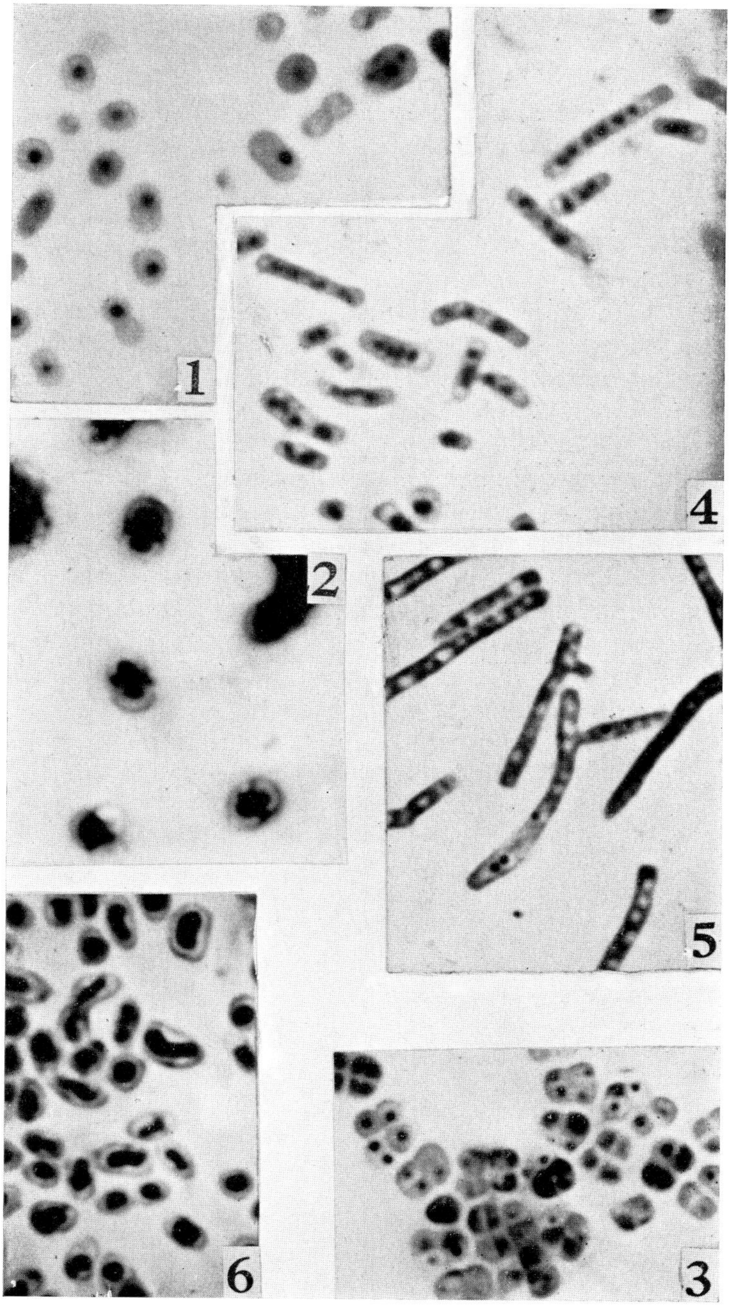

FIG. 41
APPEARANCES OF THE NUCLEUS

(1) Vesicular vegetative nucleus in a Gram-negative coccus. Methylene-blue-eosin.

(2) False appearance of vesicular nucleus in multicellular coccus. Actually the nuclear material of several cells is condensed centrally. Methyl-violet-nigrosin.

(3) As (1) in *Sarcina* sp.

(4), (5) Nuclear bodies in *Mycobacterium lacticola* and in *Nocardia* sp. These appear spherical, but this may be because they are too small to be resolved.

(6) Appearance of vesicular nucleus in acid-hydrolysed cells of *Aerobacter* sp., probably due to laking of the stain.

All plates at $\times 3000$.

appearance, can be distinguished in this phase of growth [44] (Fig. 40).

THE VEGETATIVE NUCLEUS

The vegetative nucleus is a concept based on studies of stained preparations, made with the classical light microscope [1-4, 15, 20, 44, 45] and confirmed by phase-contrast microscopy in living bacteria. [24] No exact correlation has yet been made between these appearances and the ultra-microscopic structures demonstrable by the electron microscope in ultra-sections. [26, 27, 46]

In the young cultures of most eubacteria, myxobacteria and such chlamydobacteria and caulobacteria as have been described, as well as in the primary mycelium of actinomyces and streptomyces, the nucleus consists of short rod-like bodies, sometimes slightly broader at the ends than the centre, lying transversely to the long axis of the bacterium, and occupying almost its entire width. Usually, they lie in pairs, but these are not analogous with chromosome pairs, and their exact identity with the chromosomes of plant and animal cells is dubious. The pairs may lie parallel to one another or at a slight angle. They are sometimes so close together as not to be resoluble separately by the microscope, and sometimes quite widely separated.

Although the bodies appear rod-shaped, they show a marked tendency to present the long axis of the rod to the observer. It has been suggested that they are, in fact, in the form of a short, spiral band [31] and this is to some extent supported by the observed form of the nuclei of *Caryophanon latum*, which resembles the eubacteria in many of its morphological attributes, but is much larger. [47, 48] Its nuclei are ring or disc-shaped, and lie, as a rule, in a plane transverse to the long axis of the bacillus-like organism (Fig. 5). However, a clear demonstration of a primary vegetative nucleus in the form of unequivocal transverse rods is given by Oscillospira, another, similar micro-organism that is even larger than Caryophanon [49] (Fig. 44).

The slightly dumbbell-shaped appearance of the nucleoids is frequently seen, and phase-

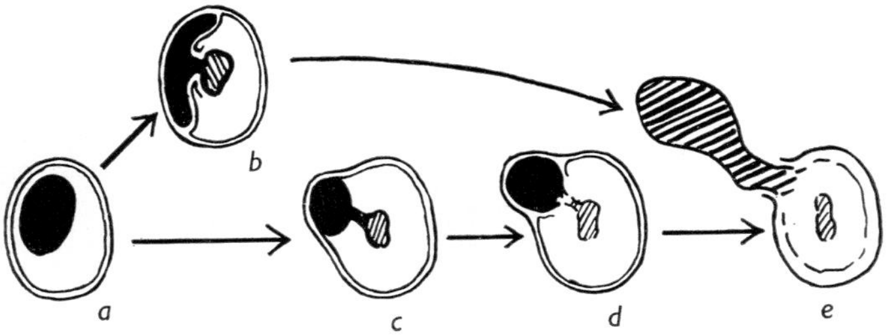

(*Reproduced from the Journal of Hygiene*)

FIG. 42
EFFECTS OF HYDROLYSIS ON THE SPORE NUCLEUS

a. Spore with nucleus in natural position.

b. The 'crescentic nucleus'. The nuclear material forms a pool between the cytoplasm and spore coat.

c, *d*. The 'peripheral nucleus'. The spore coat is bulged outwards by the ejected nuclear material.

e. Complete ejection of nuclear material.

contrast studies seem to show that it is indicative of division in the transverse plane, [24] although division of true chromosomes is invariably by longitudinal splitting. It is obvious that if bacterial nuclei carry the genes of the cell, arranged in a linear manner, as from genetical considerations must necessarily be true, then their division cannot take place in any other way than by splitting, and indeed, the longitudinal division of the DNA thread in the bacterial nucleus has been demonstrated by autoradiography. [28, 29] Other schemes of nuclear division in bacteria [30, 31] have been devised to accommodate this necessity (Fig. 46).

The nuclear unit of the vegetative cell is a single nucleoid or a recently-divided pair of sister nucleoids. In bacteria of unicellular, S morphology a pair is disposed at each end of the cell, but it is probable that both pairs are of identical genetical constitution, and this concept of a reductionally-dividing, simple nucleus has been borne out by genetical studies [50] and by ultra-sectioning. [51] There appears to be only one molecule of DNA per nucleus, and replication, in the vegetative stages, goes on continuously. [51] In very young cultures of such bacteria the cells often contain only a single pair of nucleoids, and the bacteria may contain from one to four or six cells. Each of the cells of a septate bacillus, such as a sporing bacillus or R variant, contains one nucleoid or recently-divided pair, but the two morphological types are perfectly distinct (Figs 5 & 8). It is probable that these multicellular non-sporing eubacteria occur mainly in the process of germination.

The germination of the spore, microcyst or resting cell is accomplished in approximately the same manner in each case (Fig. 45). The cell commences to elongate, and in the case of some spores and microcysts, casts the outside wall. The vesicular nucleus is transformed into a single, large transverse rod, which almost immediately divides into two, and afterwards, in the case of smooth types, into four. Normal cell division then commences. [1-22] During the process of germination the cell increases in size, except in the case of myxobacteria and some Gram-negative rods, which tend rather to diminish; the nuclear material becomes large and readily stainable. This is the period of the lag phase of the culture. In the logarithmic phase, which immediately follows, the bacteria at first divide by simple fission or budding alone, but later this method is accompanied or superseded by others more complex, and eventually the resting condition is restored. The maturation of the nucleus of the resting cell is discussed at length in a later section.

THE NATURE OF THE BACTERIAL NUCLEOID

It is not certain that the bacterial chromosome exactly corresponds to those of other cells. It performs the function of ensuring the correct distribution of genes, which are now known to be arranged upon it in a linear manner; and although in this particular it resembles other chromosomes, it differs from them in a number of respects. It is actually less susceptible than the resting nucleus to radiations [52] and being simple it does not require an elaborate mitotic process to ensure correct distribution on cell division. Evidence of a cyclic change from a tape-like to a more compact form of chromosome has been presented [53, 54, 55] but the significance of this is not at all clear. It seems likely that the coiling of the DNA thread is less regular and complex in bacterial nucleoids than in other types of chromosome.

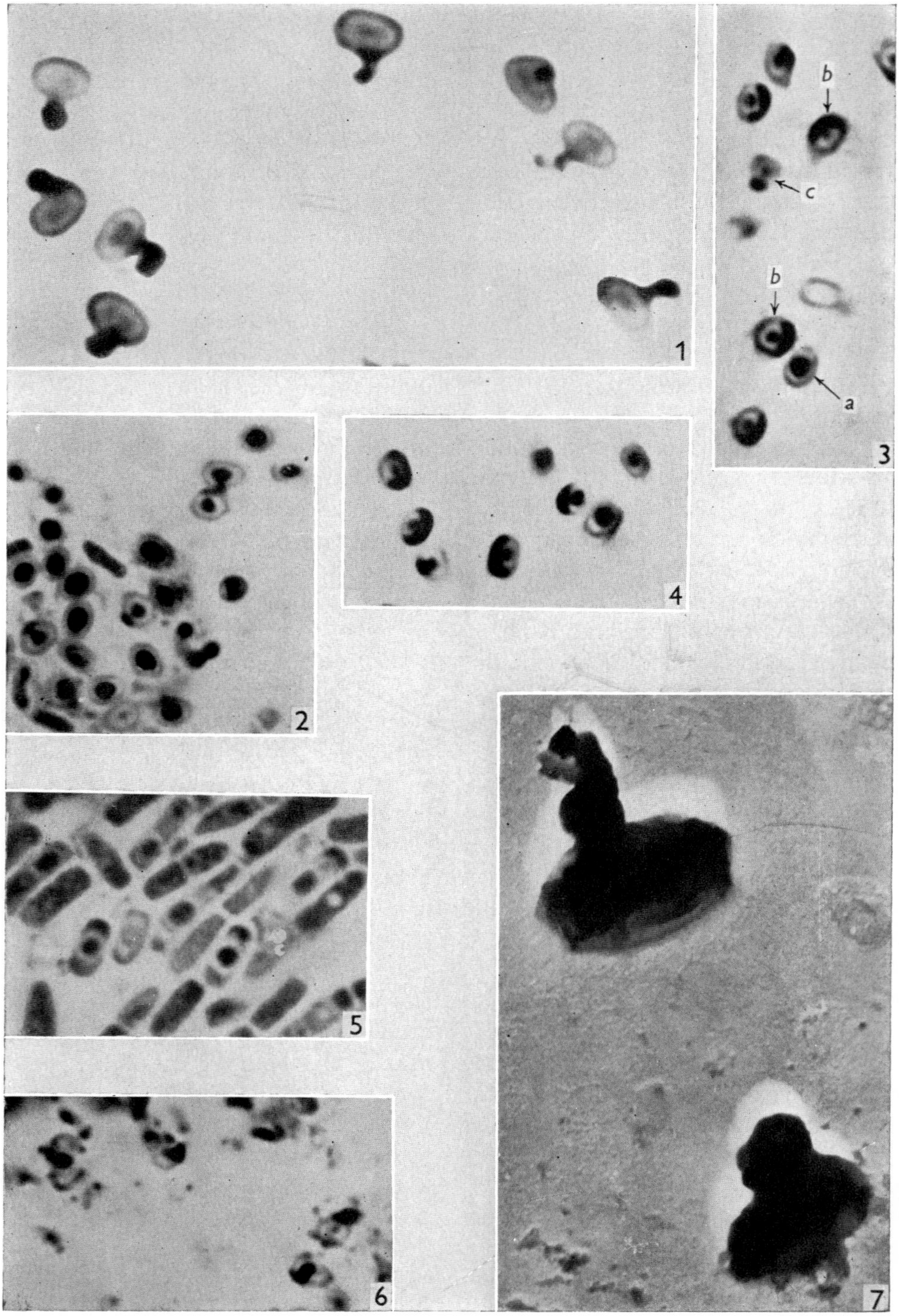

(*Reproduced from the Journal of Hygiene*)

Early genetic studies upon certain bacteria led to the adoption of theories requiring the existence of three chromosomes or even of a branched chromosome; but more recently it has been established that genetically, as well as cytologically, a single chromosome must be assumed, at least in those cases, notably *Escherichia coli*, that have been fully investigated.

In the process of fission it appears that each member of each pair of chromosomes is identical with the others, and is in fact derived from the same parent chromosome in previous cell generations; [50, 56] unlike the condition in plant or animal cells, where the two members of the pair may be of entirely different constitution with respect to several genes. Similarly, when a smooth bacterium divides, the nucleoids at one end of the cell pass together into one daughter cell. Thus the entire nuclear complement of a bacterial cell, such as those in Plate 1 of Figure 5, with four nucleoids, is derived, at a remove of two cell divisions, from a single grandparent chromosome. While the truth of this interpretation cannot be established with certainty until technical methods permit the accurate identification of individual chromosomes, or else enable the process of nuclear division to be followed in the living cell, it is certain that the appearance of fixed and stained material, at different stages of the process, does not encourage any other interpretation. It may be argued that this method of study is unreliable, but it was until recently the only method by which the nuclear processes accompanying cell division could be studied in all types of cell; and such evidence as is available from other sources, notably phase-contrast and ultra-sectioning, is in reasonable accord with it. Genetical evidence is in complete accord, although, as has already been noted, it is limited to a very few species.

It follows, therefore, that although bacteria may commence their life, in a particular culture, with a set of two or four chromosomes of different genetic constitution, possibly derived from the sexual processes which

---

FIG. 43
THE SPORE NUCLEUS

The bacterial endospore has a vesicular resting nucleus, typical of such nuclei, except that it appears to be in a condition of turgor when mature. Under the influence of acid-hydrolysis processes the nuclear material may be partly or completely ejected, giving the various appearances which have, in the past, been described as 'peripheral' or 'crescentic' nuclei. The immature spore nucleus does not behave in this fashion. The weak spot, through which the nuclear material may be ejected, possibly represents a germination pore.

(1) Spores of *Bacillus* sp., treated with N/1 nitric acid for 5 minutes, stained Giemsa and re-stained tannic-acid-violet, to demonstrate that the ejected nuclear material is outside the spore coat. One spore has retained it within the spore coat and shows the 'peripheral nucleus'. ×5000.

(2) Spores of *B. subtilis*, showing the spore nucleus in its natural condition. Acid-Giemsa. ×3000.

(3, 4) Spores of *Clostridium welchii*. Nitric acid for 10 minutes, stained crystal violet. All types of appearances seen; *a*, nucleus in natural condition; *b*, 'crescentic nucleus'; *c*, 'peripheral nucleus'. Other spores are in intermediate stages. ×3000.

(5) Maturing spores of *Cl. welchii*, as in (3, 4), showing no change of position of nucleus.

(6) As (1), after several hours hydrolysis.

(7) As (1), electron micrograph, gold-shadowed. ×16,000.

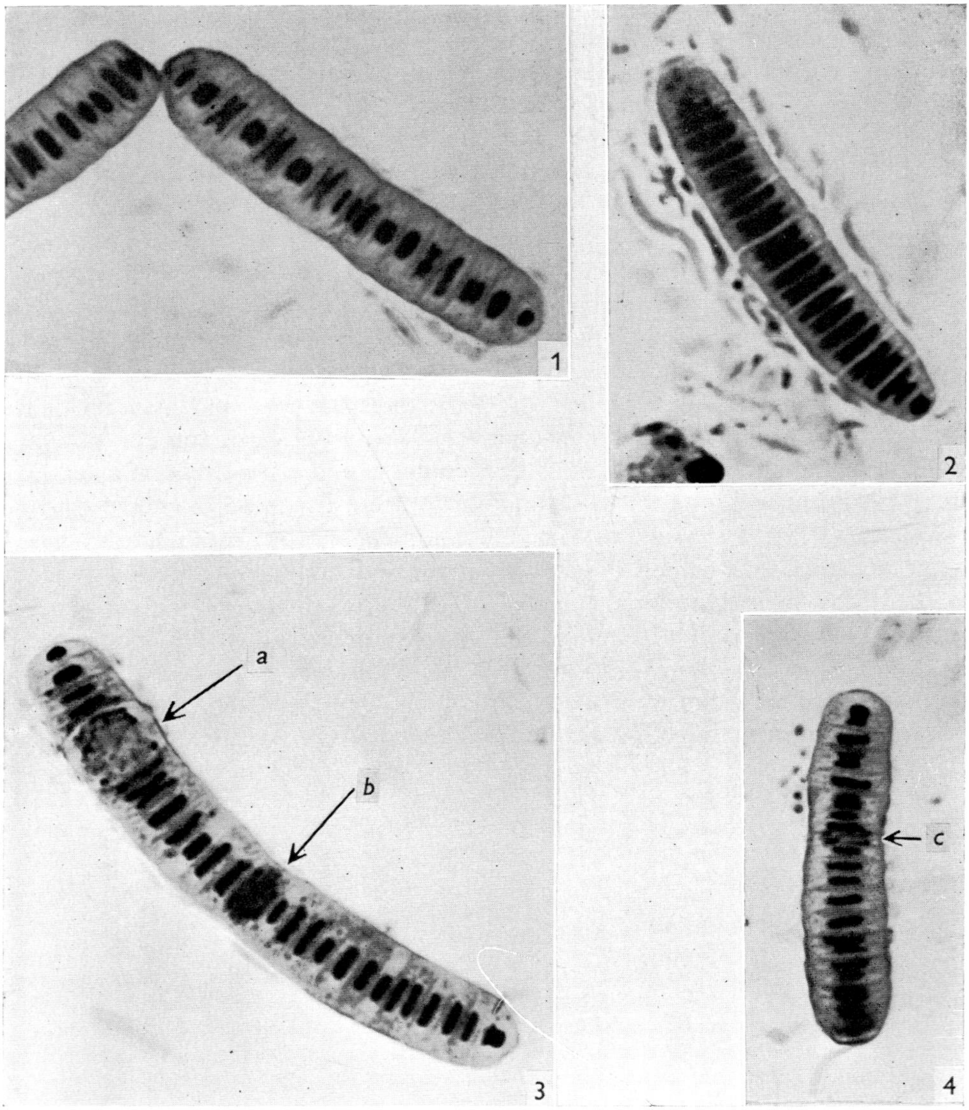

(*Reproduced from the Journal of General Microbiology by permission of Mr A. A. Tuffery*)

FIG. 44
CYTOLOGY OF OSCILLOSPIRA

Because of its relatively great size, an exceptionally clear picture of the bacterial nucleus is afforded by the giant bacterium *Oscillospira*. In (1) the twin, rod-shaped nucleoids are shown both in profile and endwise. (2) shows their arrangement within small, disc-like cells. In (3) and (4), *a*, *b* and *c* represent progressively earlier stages in the maturation of the spore and condensation of its nuclear content. Acid-Giemsa, $\times 1000$.

appear to precede the formation of the resting nucleus, these differently constituted chromosomes will rapidly become segregated, without the intervention of further sexual conjugation. It appears that the vegetative generations of bacteria of this type are haploid, but sometimes multinucleate. This conclusion is also borne out by genetic studies.[57]

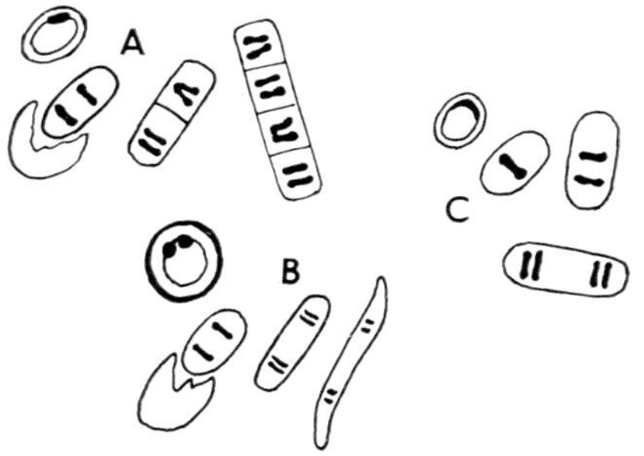

FIG. 45
THE GERMINATION OF THE RESTING STAGE

A. The spore of a rough bacillus.
B. The microcyst of *Cytophaga* sp.
C. The resting cell of *E. coli*.

The process is similar in each case. The wall may be shed or absorbed. The vesicular nucleus is transformed into one or more bar-shaped bodies and these divide to give the vegetative nucleus.

The multinucleate conditon is irregular, bacteria in the same culture may have one, two, four or more chromosome-like nuclei. It was shown by Witkin [50] that if bacteria have two or four chromosomes, then sectored colonies, with half or quarter variant sectors respectively, are produced by irradiation (Chap. 10). The segregation of the variant character, produced by alteration of an irradiated chromosome, takes place after a delay of two vegetative divisions, representing the process of segregation of the nucleus.

It has already been pointed out, in the historical section at the beginning of this chapter, that a detailed model of the chromosome and its mode of division is still rather difficult to construct. It is too near the borders of resolution by visible light, and too hard to interpret from ultra-sections in the electron microscope. Working from visible-light observations, two reasonably similar schemes were suggested by Fitz-James [30] and by Tulasne and Vendrély. [31] They envisage curved or semi-circular chromosomes, with enlarged tips, or with a third, central body also, and dividing with a rotatory movement around the axis of the cell. The two schemes are not identical, but are sufficiently alike to make it obvious that these independent workers were describing the same thing (Fig. 46). Kellenberger, [25] basing his interpretation upon sections, has described two bundles, pleated back and forth, and consisting of uncoiled DNA threads. Fuhs [51] believes that these two bundles are twined one around the other. On the other hand, Giesbrecht [58, 59, 60] interprets the appearances seen in some sections of nuclei as spiral chromosomes of a much more conventional type, with a major coil of 1000 Å period, a secondary spiral of 300 Å, and a molecular-sized spiral of 70 Å. Since it is excessively difficult to fix, embed and section bacteria without causing gross distortion of the nucleus, it is quite possible that this complex structure really exists, in undisturbed material (Fig. 37).

SYNGAMOUS VEGETATIVE REPRODUCTION

This mode of reproduction is common in many types of bacteria, but does not appear in spore-bearing eubacteria. [12, 15, 19, 33, 62, 63] The nuclear appearances which accompany it are very striking, [61, 63] and it is curious that the process has attracted very little atten-

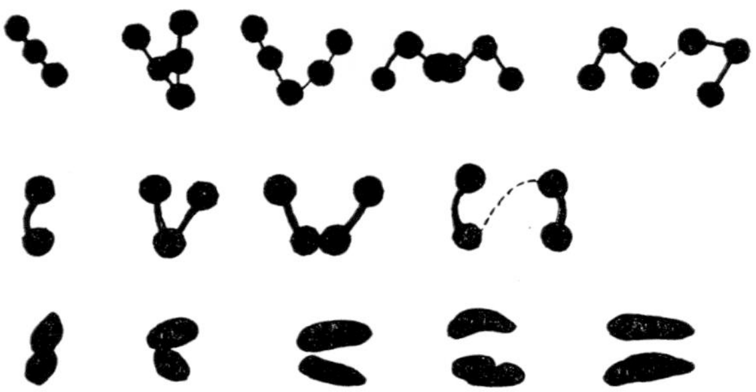

FIG. 46
DIAGRAMS OF SCHEMES OF NUCLEAR DIVISION

A, after Fitz-James, 1954; B, after Tulasne & Vendrély, 1954; both reconstructed from stained preparations. C, after Mason & Powelson, 1955, as visualised by time-lapse phase-contrast photomicrography.

In each case, the nucleoids are seen to separate with a rotatory movement, but the living material conveys the impression of transverse, rather than longitudinal fission.

tion from geneticists. It consists essentially of a nuclear fusion in a cell, which grows into a short filament and then divides or fragments into a new generation of bacilli of normal length (Figs 47, 48 & 49). In contrast to the filaments of rough cultures, which are multicellular, repeating in each unit of the chain the nuclear pattern of the individual, the reproductive filaments are unicellular and their nuclear material is arranged in a distinctive manner. Such filaments may, however, occur in cultures of both smooth and rough morphology, and even in cocci. Although there are recognisable differences between the various types, the general plan is similar in all of them (Fig. 50).

In smooth cultures, the shortest filaments, which are about twice the length of a single bacterium, have their nuclei packed together at the centre of the cell. [61] Where they can be distinguished separately they are invariably found to consist of three pairs. Filaments of slightly greater length contain six pairs of bodies, which may be together at the centre of the cell, or at a later stage, distributed in pairs throughout the length of the filament. This increase represents a single nuclear division within the fusion cell. A second nuclear division occurs after the pairs have been redistributed in the filament, and the latter then fragments into individual bacteria, each containing two pairs. Thus each nucleus of the original six becomes the parent of the entire complement of two pairs in one daughter bacterium. [61] The occurrence of the fusion process has now been confirmed by genetic studies [50] since Witkin has shown that filaments of this type behave as if they possessed only a single nucleus, in irradiation experiments, but nothing has been recorded of the process whereby the fusion cell attains its form and nuclear complement. Bacteria occur that are probably the precursors of the fusion cells, as they contain three pairs of nuclei, arranged in a more normal manner, with one pair at each end and one in the centre of the bacterium. But how these are derived from a bacterium with two pairs is not easy to understand, and some undetected process of reduction may be entailed (Fig. 47).

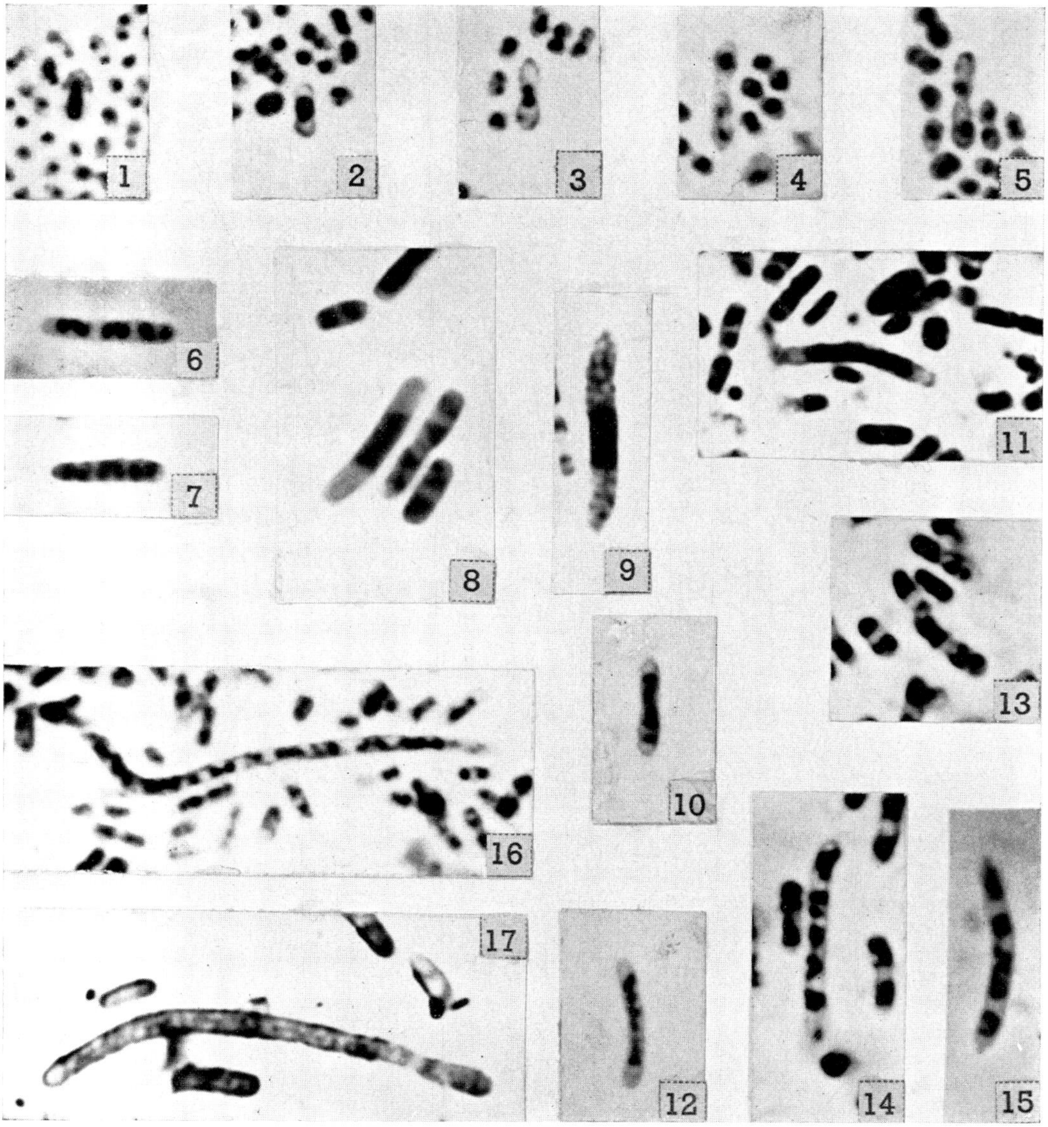

(*Reproduced from the Journal of Hygiene*)

FIG. 47
COMPLEX VEGETATIVE REPRODUCTION

(1-5) *Streptococcus faecalis*, ×3000; (6-17) *Shigella flexneri*, ×3000.
(1) Large cell with elongated fusion nucleus.
(2-4) Development of filament.
(5) Fragmentation of filament.
(6, 7) Trinucleate pre-fusion cells.
(8, 9) Fusion nuclei. In (8) compare fusion cell (*left*) with dividing cell (*centre*) and normal cell (*right*).
(10, 11, 12) Development of fusion nucleus.
(13, 14, 15) Redistribution of nuclear elements in growing filament.
(16, 17) Fragmentation of filament. (16) stained for nuclear structures, (17) for cell walls.

This method of reproduction, as it occurs in rough bacteria and also in streptococci, is essentially similar to that in smooth bacteria but differs slightly in detail. The fusion cell of a lactobacillus is approximately the same size as the four-celled rough bacterium, and contains two large nuclear bodies. [62]

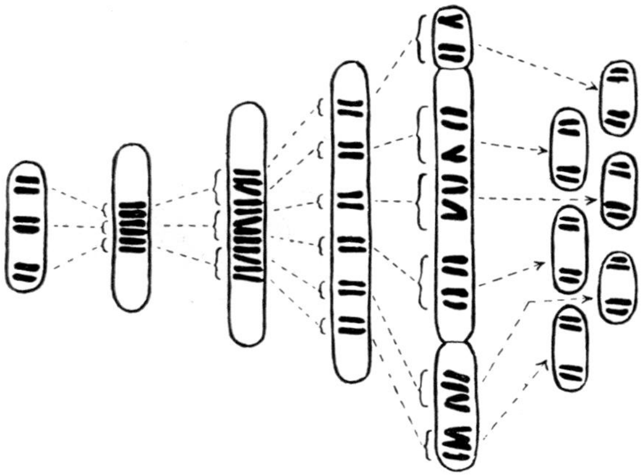

(*Reproduced from the Journal of Hygiene*)

FIG. 48
COMPLEX VEGETATIVE REPRODUCTION IN *BACTERIUM*

The trinucleate cell precedes the fusion cell with its three pairs of chromosome complexes. Within the fusion cell there is one nuclear division. The six chromosome pairs are redistributed in the growing filament. There is a second nuclear division and the filament fragments into six daughter cells, each with the full complement of two pairs of chromosome complexes. In streptococci only three daughter cells are formed.

The filament increases in length and eventually fragments, exactly as in the case of the smooth types, but the behaviour of the nuclear material is slightly different. The central mass of chromatin retains its identity for some time after the commencement of growth. Small fragments break off and migrate along the filament, and eventually the pairs are evenly distributed along its length. The filament then fragments into bacillary forms; how many has not been determined.

In the case of *Streptococcus faecalis*, which is a short-chained streptococcus, resembling a smooth bacterium in the possession of two pairs of nuclear units, the fusion nucleus is rod-shaped, with its axis longitudinally disposed in the oval coccus. The rod is transformed into a single, central mass, from which small fragments break off, as in the case of the lactobacillus.[62] The filament which is produced is comparatively short, and gives rise to three instead of six cells upon fragmentation. The occurrence, in both cases, of a multiple of three, which is the number of pairs of nuclei in the smooth type of fusion cell, indicates a similarity of constitution of the streptococcal fusion cell (Fig. 47).

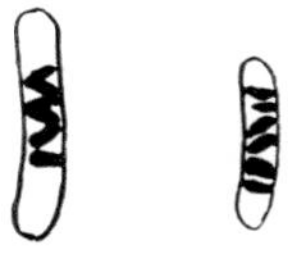

(*Reproduced from the Journal of Hygiene*)

FIG. 49
TRACINGS OF PHOTOMICROGRAPHS OF VEGETATIVE FUSION CELLS

A. *Shigella flexneri.*
B. *E. coli.*
Showing three pairs of chromosome complexes.

Filamentous cells, containing chromosome fusions of this type, occur also in myxobacteria and in chlamydobacteria, but the details of the process in these bacterial orders have not been described. [15, 33]

In addition to vegetative cells which resemble those of other, Gram-negative bacteria, cultures of Proteus contain filaments of considerable length, each having a large num-

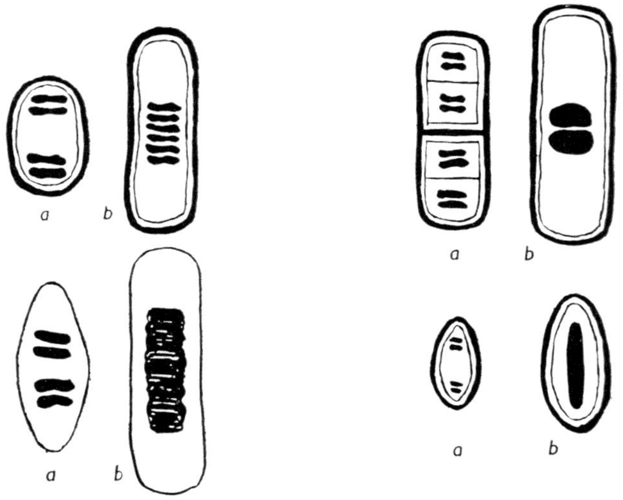

(Reproduced from the Journal of Hygiene)

FIG. 50
THE PRIMARY NUCLEUS AND VEGETATIVE FUSION CELLS IN VARIOUS BACTERIA

(*a*) Primary nucleus; (*b*) fusion cell in each case.
Top left—Smooth bacterium.
Top right—Rough bacterium.
Bottom left—Myxobacterium.
Bottom right—Short-chained streptococcus.

ber of nuclear units. These filaments appear to be unicellular, but their nuclear material is arranged in a simple, repetitive pattern, unlike that of the reproductive filaments. Their function is distributive and they comprise the swarm. Evidence has been presented to suggest that these swarmer filaments take part in a reproductive cycle, including the formation of zygospores at the points of contact of swarms. This interpretation will be discussed in Chapters 6 and 7 (Fig. 68).

Other types of filamentous cell are common in bacterial cultures. Some of these contain reduced or disorganised nuclear material. Their significance is unknown, and they may be pathological.

THE SECONDARY NUCLEUS

The secondary type of bacterial nucleus was first described in an interesting paper by Stoughton [(10)] and later, independently by Piekarski, [(2)] Bisset, [(20)] and Tulasne and Vendrély, [(31)] by all of whom it was contrasted with the primary form, but it has been little studied. This is in part explained by the fact that Stoughton failed to demonstrate the primary nucleus by the technique which he employed, so that his findings were not correlated with the later observations upon this phase of the nuclear cycle, whereas Piekarski, although successful in demonstrating both types of nucleus by the Feulgen reaction and

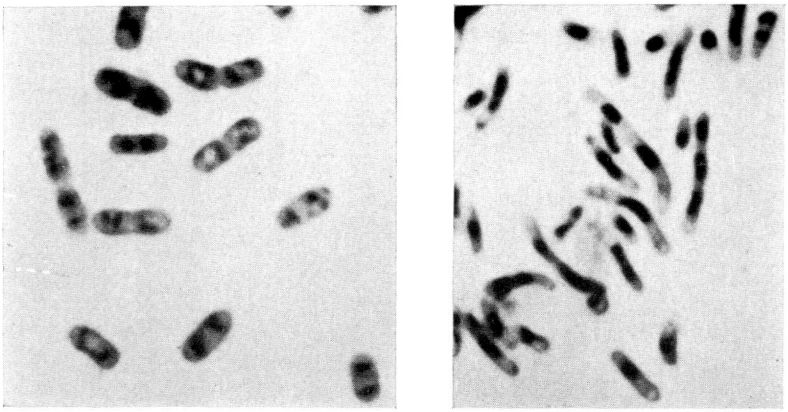

(Reproduced from the Journal of Hygiene)

FIG. 51
THE VEGETATIVE NUCLEUS IN *E. COLI*

Left—Primary form.
Right—Secondary form.
Acid-Giemsa ×3000.

by the acid-Giemsa technique as well as by ultra-violet light, failed to resolve them properly, and figured them as spherical bodies. This cast doubt upon the value of his morphological interpretations when the true form of the primary nucleus was afterwards discovered. In fact the morphology of the secondary nucleus is considerably less easy than that of the primary nucleus to define. It consists of a structure which, although it may appear single is probably always paired and is disposed centrally in the bacillus.[20] It does not stain with the same clarity as the primary nucleus, so that its exact form is difficult to determine. It divides with the cell, and has been observed to do so by dark-ground illumination (Figs 51 & 52).

The secondary nucleus is an alternative form that may or may not be adopted in the later vegetative stages of a culture. If it is not adopted the nucleus may take the form of a central, chromatinic rod, and thereafter proceed to the formation of the resting nucleus.[20] It has not been recorded as occurring in spore-bearing genera. These pass directly from the primary nuclear phase to the changes associated with sporulation (Chap. 6).

The secondary nucleus was described by Piekarski as a single, spherical nucleus situated at the centre of the bacterium, in contrast to the primary nuclei, which he described as similar, spherical bodies disposed at each end, in a smooth culture of *Salmonella typhi*. Stoughton described it as a single or double structure, not unlike a pair of primary chromosomes, although broader. The organism which he studied, a plant pathogen, formed the resting stage, a spherical body, by a rather unusual method from the secondary nucleus (Chap. 6). By Stoughton's technique, a vital staining method, using carbol fuchsin, young cultures of this bacterium stained uniformly, whereas the nuclear structures of older cultures were clearly visible. This was almost certainly due to the masking effect of the cell membrane in the young cultures (Chap. 2).

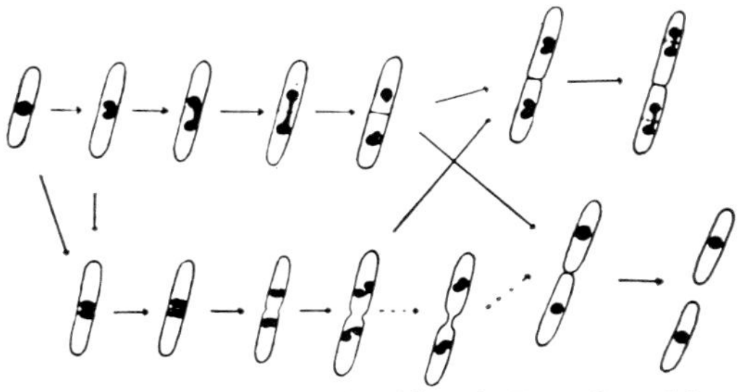

(After Stoughton. Reproduced from the Proceedings of the Royal Society)

FIG. 52
THE SECONDARY NUCLEAR PHASE

Postulated alternative methods of division in *Bacterium malvacearum*.

Bacteria in the secondary nuclear phase often produce forms that appear to be in process of sexual conjugation. They are attached end to end, usually at a slight angle, with the nuclear material concentrated at the point of contact. From such forms Strough-ton described the formation of a spherical body resembling the microcyst of myxobacteria. When the resting nucleus is produced from similar forms in *E. coli* the process differs from that found in Stoughton's bacterium, which he describes as extruding the spherical body from the point of conjugation, or from the side of a bacterium. This phenomenon will be discussed in the section on the formation of the resting nucleus (Figs 36, 57 & 68).

The most probable explanation of the appearance of the secondary nucleus is that it represents a phase of reduced nuclear activity in the latter stages of culture.

THE ROD-LIKE NUCLEUS

The bacterial nucleus sometimes takes the form of a straight or spiral rod, lying parallel with the long axis of the cell (Figs 53, 54 & 60). This appearance has very often been described, and doubt has also been cast upon its validity. Lewis [64] expressed the opinion that it is an optical illusion caused by viewing the meshwork of stainable cytoplasm left in a cell containing an accumulation of unstained granules of fat or other reserve materials. Delaporte [65, 66] showed that similar appearances can result from the action of various extraneous agencies, such as aeration, but that they can also be normal and apparently genuine, especially in the process of maturation of the resting cell. In fact, wherever these processes have been studied, this form of nucleus has been claimed to occur [12-16, 67, 68] as a stage in the formation of the nucleus of every kind of spore and cyst produced by bacteria. The most recent work, on the electron microscopy of ultra-sections, has confirmed the claims of earlier and later workers with the light microscope [54, 69, 70, 71] going back to Schaudinn [72] at the beginning of the century.

Since the existence of this type of nucleus cannot easily be denied, it remains to explain its nature. Schaudinn believed it to be a condensation of small granules. Bisset [73] suggested that it is composed of a pile of small primary nuclei, like coins, and that it represents a polyploid condition. Thus it is capable of simple fission, like a protozoan meganucleus. Some known diploid nuclei have this appearance (Fig. 54).

Bacterial cells in this nuclear phase are still

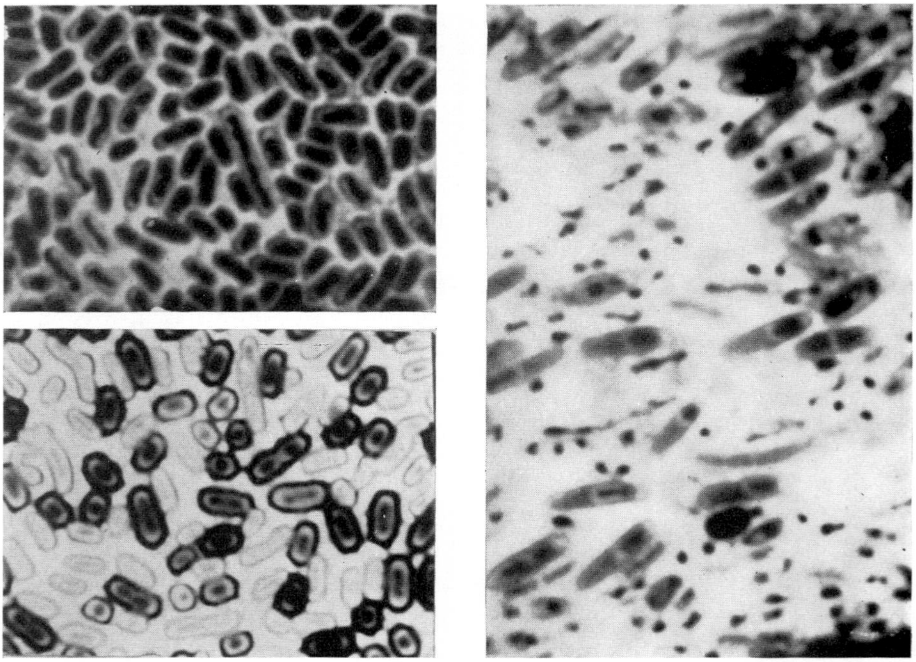

(Reproduced from the Journal of Hygiene and Journal of General Microbiology)

FIG. 53
TYPES OF ROD-LIKE NUCLEUS

Top left—Early stages in the maturation of the resting nucleus in *E. coli.*
Bottom left—Similar stages in *A. aerogenes* (the dark surface material is capsular)
Right—Rod-forms enclosed in the developing spore of *Cl. tetani.* Acid-Giemsa ×3000.

capable of vegetative reproduction. Gram-negative, intestinal bacteria may adopt the condition in cultures not more than a day old, although they do not invariably do so, as it is alternative to the secondary nuclear phase, which is often found in such circumstances.[20]

Appearances suggestive of a rod nucleus may sometimes be due simply to an accidental arrangement of amorphous material within a rod-shaped bacillus, but the nucleus is nevertheless capable of retaining the rod form when enclosed in an oval spore during maturation [14, 65] so that it must have a degree of coherence (Fig. 53).

The entire secondary mycelium of streptomyces, from which the spore-bearing hyphae arise, contains rod-shaped nuclei, and they are frequently found in the early stages of sporulation in Bacillus and Clostridium, and it is probable that, in this case also, vegetative reproduction is still in progress (Fig. 54).

NUCLEAR FISSION

Robinow's [4] clearly-illustrated concept of the bacterial nucleus remains generally valid for optical demonstrations, whether stained or by phase-contrast of the living material. It

REFERENCES

1. STILLE, B. (1937). *Arch. Mikrobiol.* **8**, 125.
2. PIEKARSKI, G. (1937). *Arch. Mikrobiol.* **8**, 428.
3. GUTSTEIN, M. (1924). *Zentbl. Bakt. ParasitKde I, Orig.* **93**, 393.
4. ROBINOW, C. F. (1942). *Proc. R. Soc.* **B**, **130**, 299.
5. PAILLOT, A. (1919). *Annls Inst. Pasteur, Paris*, **33**, 403.
6. BADIAN, J. (1930). *Acta Soc. Bot. Pol.* **7**, 55.
7. BADIAN, J. (1933). *Arch. Mikrobiol.* **4**, 409.
8. BADIAN, J. (1933). *Acta Soc. Bot. Pol.* **10**, 361.
9. KRZEMIENIEWSKA, H. (1930). *Acta Soc. Bot. Pol.* **7**, 507.
10. STOUGHTON, R. H. (1929). *Proc. R. Soc.* **B**, **105**, 469.
11. STOUGHTON, R. H. (1932). *Proc. R. Soc.* **B**, **111**, 46.
12. KLIENEBERGER-NOBEL, E. (1947). *J. gen. Microbiol.* **1**, 33.
13. KLIENEBERGER-NOBEL, E. (1945). *J. Hyg., Camb.* **44**, 99.
14. BISSET, K. A. (1950). *J. gen. Microbiol.* **4**, 1.
15. GRACE, J. B. (1951). *J. gen. Microbiol.* **5**, 519.
16. FLEWETT, T. H. (1948). *J. gen Microbiol.* **2**, 325.
17. KLIENEBERGER-NOBEL, E. (1947). *J. gen. Microbiol.* **1**, 22.
18. MORRIS, E. O. (1951). *J. Hyg., Camb.* **49**, 46.
19. MORRIS, E. O. (1951). *J. Hyg., Camb.* **49**, 175.
20. BISSET, K. A. (1949). *J. Hyg., Camb.* **47**, 182.
21. BISSET, K. A. (1951). *Cold Spring Harb. Symp. quant. Biol.* **16**, 373.
22. BISSET, K. A., GRACE, J. B. & MORRIS, E. O. (1951). *Expl Cell Res.* **2**, 388.
23. PULVERTAFT, R. J. V. (1950). *J. gen. Microbiol.* **4**, 14.
24. MASON, D. J. & POWELSON, D. M. (1956). *J. Bact.* **71**, 474.
25. KELLENBERGER, E. (1960). *Symp. Soc. gen. Microbiol.* **10**, 39.
26. VAN ITERSON, W. & ROBINOW, C. F. (1961). *J. biophys. biochem. Cytol.* **9**, 171.
27. GIESBRECHT, P. (1962). *Zentbl. Bakt. ParasitKde I, Orig.* **187**, 452.
28. CAIRNS, J. (1963). *Cold Spring Harb. Symp. quant. Biol.* **28**, 43.
29. CAIRNS, J. (1963). *J. molec. Biol.* **6**, 208.
30. FITZ-JAMES, P. C. (1954). *J. Bact.* **68**, 464.
31. TULASNE, R. & VENDRÉLY, COLETTE (1954). *Schweiz. Z. allg. Path. Bakt.* **17**, 49.
32. POCHON, J., TCHAN, Y. T. & WANG, T. L. (1948). *Annls Inst. Pasteur, Paris*, **74**, 182.
33. BISSET, K. A. (1948). *J. Hyg., Camb.* **46**, 264.
34. BISSET, K. A. (1949). *J. gen. Microbiol.* **3**, 93.
35. BISSET, K. A. & HALE, C. M. F. (1953). *J. gen. Microbiol.* **8**, 442.
36. CLARK, J. B. & WEBB, R. B. (1953). *J. Bact.* **66**, 498.
37. BISSET, K. A. (1950). *Expl Cell Res*, **1**, 473.
38. BISSET, K. A. (1950). *Nature, Lond.* **166**, 431.
39. BRADLEY, D. E. & WILLIAMS, D. J. (1957). *J. gen. Microbiol.* **17**, 75.
40. BISSET, K. A. & HALE, C. M. F. (1951). *J. Hyg. Camb.* **49**, 202.
41. ROBINOW, C. F. (1951). *J. gen. Microbiol.* **5**, 439.
42. ROBINOW, C. F. (1953). *J. Bact.* **66**, 300.
43. FITZ-JAMES, P. C. & YOUNG, E. (1959). *J. Bact.* **78**, 755.
44. ROBINOW, C. F. (1945). Addendum to *The Bacterial Cell*, ed. Dubos, R. J., Harvard University Press.
45. ROBINOW, C. F. (1956). *Bact. Rev.* **20**, 207.
46. GLAUERT, A. (1962). *Br. med. Bull.* **18**, 245.
47. PRINGSHEIM, E. G. & ROBINOW, C. F. (1947). *J. gen. Microbiol.* **1**, 267.
48. TUFFERY, A. A. (1955). *Expl Cell Res.* **9**, 182.
49. TUFFERY, A. A. (1954). *J. gen Microbiol.* **10**, 342.
50. WITKIN, E. M. (1951). *Cold Spring Harb. Symp. quant. Biol.* **16**, 357.
51. FUHS, G. W. (1965). *Bact. Rev.* **29**, 277.
52. RUBIN, B. A. (1954). *J. Bact.* **67**, 361.
53. WHITFIELD, J. F. & MURRAY, R. G. E. (1956). *Can. J. Microbiol.* **2**, 245.
54. PREUSSER, H. J. (1958). *Arch. Mikrobiol.* **29**, 17.
55. PREUSSER, H. J. (1959). *Arch. Mikrobiol.* **33**, 105.
56. RYAN, F. J. & WAINWRIGHT, J. K. (1954). *J. gen. Microbiol.* **11**, 364.
57. LEDERBERG, J. (1948). *Heredity, Lond.* **2**, 145.
58. GIESBRECHT, P. & PIEKARSKI, G. (1958). *Arch. Mikrobiol.* **31**, 68.
59. GIESBRECHT, P. (1959). *Zentbl. Bakt. ParasitKde, I, Orig.* **176**, 413.

60. GIESBRECHT, P. (1961). *Zentbl. Bakt. ParasitKde, I, Orig.* **183**, 1.
61. BISSET, K. A. (1948). *J. Hyg., Camb.* **46**, 173.
62. BISSET, K. A. (1948). *J. gen. Microbiol.* **2**, 248.
63. ROBINOW, C. F. (1944). *J. Hyg., Camb.* **43**, 143.
64. LEWIS, I. M. (1941). *Bact. Rev.* **5**, 181.
65. DELAPORTE, B. (1950). *Adv. Genet.* **3**, 1.
66. DELAPORTE, B. (1956). *Bull. Soc. bot. Fr.* **103**, 521.
67. YOUNG, I. E. & FITZ-JAMES, P. C. (1959). *J. Biophys. Biochem. Cytol.* **6**, 467.
68. YOUNG, I. E. & FITZ-JAMES, P. C. (1959). *J. Biophys. Biochem. Cytol.* **6**, 483.
69. RYTER, A. & JACOB, F. (1964). *Annls Inst. Pasteur, Paris*, **107**, 384.
70. ELLAR, D. J. & LUNDGREN, D. G. (1966). *J. Bact.* **92**, 1748.
71. GRULA, E. A., SMITH, G. L. & GRULA, M. M. (1968). *Can. J. Microbiol.* **14**, 293.
72. SCHAUDINN, F. (1903). *Arch. Protistenk*, **2**, 421.
73. BISSET, K. A. (1956). *Symp. Soc. gen. Microbiol.* **6**, 1.
74. JACOB, F. & BRENNER, S. (1963). *C. r. hebd. Séanc. Acad. Sci. Paris*, **256**, 298.
75. JACOB, F., BRENNER, S. & CUZIN, F. (1963). *Cold Spring Harb. Symp. quant. Biol.* **28**, 329.
76. JEYNES, M. H. (1961). *Expl Cell Res.* **24**, 255.
77. BERGERSEN, F. J. (1953). *J. gen. Microbiol.* **9**, 26.
78. RYTER, A. & JACOB, F. (1966). *Annls Inst. Pasteur, Paris*, **110**, 801.
79. STANIER, R. Y. & VAN NIEL, C. B. (1962). *Arch. Mikrobiol*, **42**, 17.
80. BISSET, K. A. (1959). *Vistas Bot.* **1**, 313.
81. LWOFF, A. (1950). *Problems of Morphogenesis in Ciliates*. New York: Wiley.
82. FREEMAN, J. A. (1964). *Cellular Fine Structure*. London: McGraw-Hill.
83. BISSET, K. A. & HALE, C. M. F. (1951). *J. gen. Microbiol.* **5**, 150.
84. GRACE, J. B. (1954). *J. gen. Microbiol.* **10**, 325.
85. ELLAR, D. J., LUNDGREN, D. G. & SLEPECKY, R. A. (1967). *J. Bact.* **94**, 1189.
86. FITZ-JAMES, P. C. (1965). *Symp. Soc. gen. Microbiol.* **15**, 369.
87. GLAUERT, A. M., KERRIDGE, D. & HORNE, R. W. (1963). *J. Cell Biol.* **18**, 327.
88. ABRAM, D., VATTER, A. E. & KOFFLER, H. (1966). *J. Bact.* **91**, 2045.
89. REMSEN, C. C., WATSON, S. W., WATERBURY, J. B. & TRÜPER, H. G. (1968). *J. Bact.* **95**, 2374.
90. FITZ-JAMES, P. C. (1964). *J. Bact.* **87**, 1483.
91. JEYNES, M. H. (1955). *Nature, Lond.* **176**, 1077.
92. BAYER, M. E. (1968). *J. gen. Microbiol.* **53**, 395.
93. FUHS, G. W. (1969). *The Nuclear Structures of Protocaryotic Organisms*. New York: Springer.

CHAPTER 5

Autogamous and Sexual Processes

HISTORICAL

The idea of sexuality in bacteria has had a rather curious history. As the existence of the nucleus itself was denied for many years, in the face of a good deal of evidence, it is not surprising that sexual conjugation was regarded as even more improbable, although cytological processes that closely resemble the types of conjugation found in other protista have been known for a long time. Mellon [1] published photomicrographs in 1925 that showed bacteria, apparently conjugating end-to-end, with the production of a zygospore, and shortly afterwards Stoughton [2] described the nuclear cycle accompanying such a process in a plant-pathogenic bacterium (Fig. 36). Plant pathogens are also peculiarly liable to undergo a very dramatic type of conjugation, where as many as a dozen or twenty bacteria join together by one pole in star formations (Figs 56 & 57). This has been known since the last century [3] and has been much studied by German bacteriologists [4-7] in particular, from both the cytological and genetical viewpoint.

Quite separately, bacterial cytologists have been pointing out, since the nucleus was first demonstrated, that the maturation of the endospore, spore or microcyst entails a process that has every appearance of an autogamous nuclear fusion. [8-12] The putative fusion nucleus, the longitudinal rod form, was described in the last chapter. Of this phenomenon, no genetical studies have been made. However, in 1946, at about the same time as these investigations were in progress, the interest of geneticists was aroused when Lederberg and Tatum [13] produced evidence of genetic recombination in *Escherichia coli*. This was rapidly followed by an explosion of research upon the genetics of bacteria, that culminated in the elucidation of the replication of DNA, and does not require to be expounded afresh in these pages.

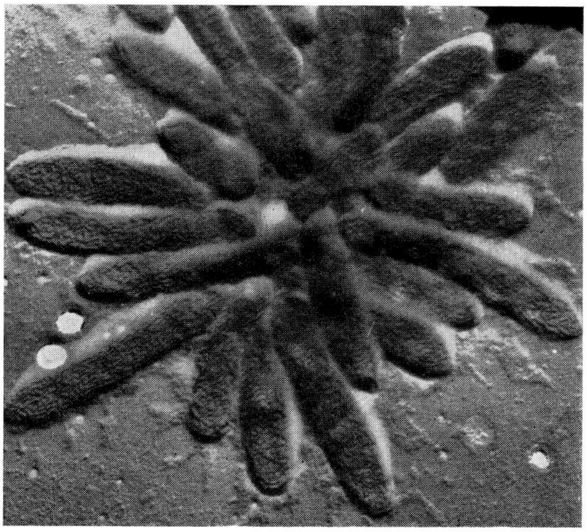

(*Electron micrograph by Dr Phyllis Pease*)

FIG. 56
STAR FORMATION
Conjugating cluster of pseudomonads. ×11,000.

Unluckily for the progress of bacterial cytology, the sexual conjugation that produces these genetical effects is a rather peculiar one, and has been observed, for practical

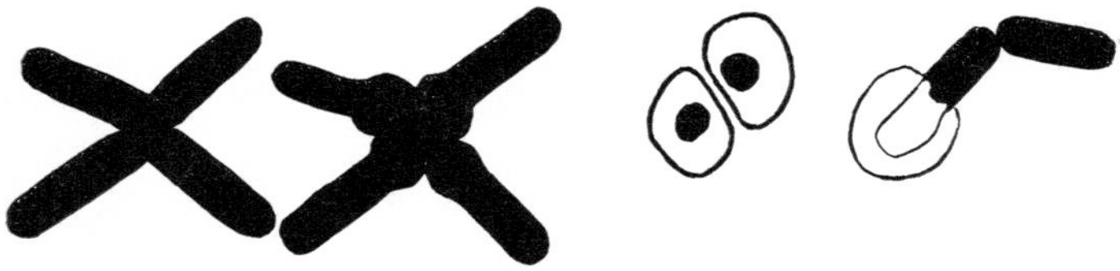

FIG. 57
STAR FORMATION

Diagram after Heumann, showing conjugation and germination in a plant-pathogenic bacterium.

purposes, almost exclusively in a very few strains of *Escherichia coli*. It has been studied so extensively and so intensively that much has been learned from it about the fundamental processes of life, but relatively little about bacteria. Indeed, under normal conditions, it seems to be rather a rare occurrence in a bacterial population, and for some time after it had been proved to exist, no visual evidence of it could be produced. This problem was solved, most ingeniously, by Lederberg [14] ten years after he had first demonstrated the genetical effect. The method, used also by Anderson, Wollman and Jacob, [15] was to bring into contact two physically dissimilar, but inter-fertile cultures, so that the conjugating pairs could be recognised microscopically. By examination of these pairs, it was found possible to demonstrate the actual protoplasmic bridge between them, in the electron microscope [15] (Figs 58 & 59).

One sexual process in bacteria has thus been proved conclusively to exist. How many others await recognition can only be a subject for conjecture, but there must be several.

NUCLEAR TRANSFER

The knowledge that certain Gram-negative bacteria can interchange genetic material by a sexual process, while in the active stage of the vegetative phase, derives from a genetical experiment. When two mutant strains of *Escherichia coli*, each of which requires to be supplied with a particular amino-acid, are mixed together in culture, hybrid strains will arise that can grow without either amino-acid. [13] In other words, they have exchanged the deficient and the complete genes that are concerned with the synthesis of these acids (Chap. 10). This occurs only when the

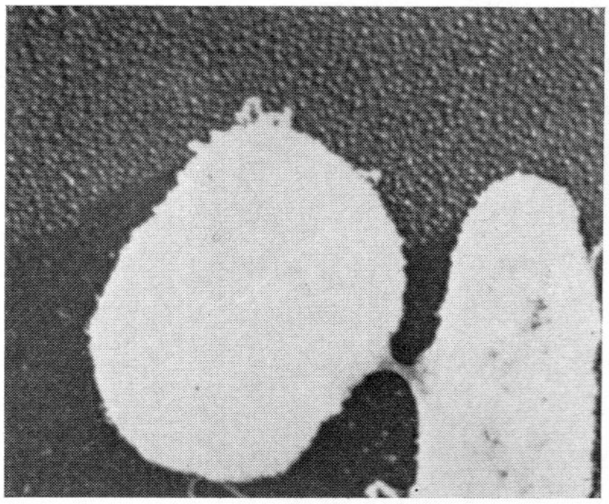

(Reproduced from the Annales de l'Institut Pasteur, Paris, by permission of Dr E. L. Wollman)

FIG. 58
CONJUGATION IN *E. COLI*

Conjugation in the vegetative phase of *Escherichia coli*. Electron micrograph. One of the two conjugating bacteria belongs to a strain with a notably long, thin morphology; the other is short and stout. It was by this device that actual contact was detected. ×20,000.

parent strains are in actual contact. The cells touch, and fuse together with the formation of a narrow protoplasmic bridge [15] (Figs 58 & 59) through which a part of the nucleus of one partner passes into the other. This, in its very limited way, is (or simulates) a true sexual process, because one strain is male or positive, and passes its nuclear material to the female or negative strain, which becomes the zygote; [16, 17, 18, 19] the donor cell is dispensable. An unusual and curious feature of this type of mating is that only a part of the male chromosome is passed over, in most cases. [20] The whole process takes about half an hour to complete, and it can be interrupted by mechanical shaking of the culture, which separates the couples at various stages in the process [21, 22] so that the linear arrangement of genes on the transferred chromosome can be studied. This aspect of the matter is discussed in Chapter 10, on the cytogenetics of bacteria.

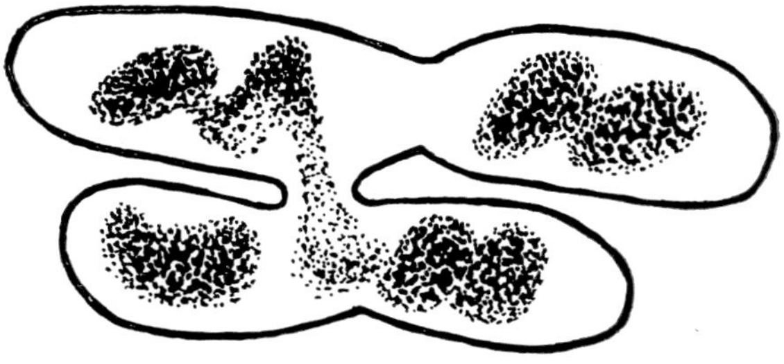

FIG. 59
CONJUGATION IN *E. COLI*

Conjugation in the vegetative phase of *Escherichia coli*. showing passage of nuclear material from one cell to another. (Drawn after an original photomicrograph kindly supplied by Professor J. Lederberg.)

The nature of the force that acts upon the chromosome, to transfer it, or part of it, from the donor to the recipient bacterium, poses an unusual problem. The solution proposed by Jacob and Brenner [23, 24] is an extension of their replicator hypothesis, discussed in

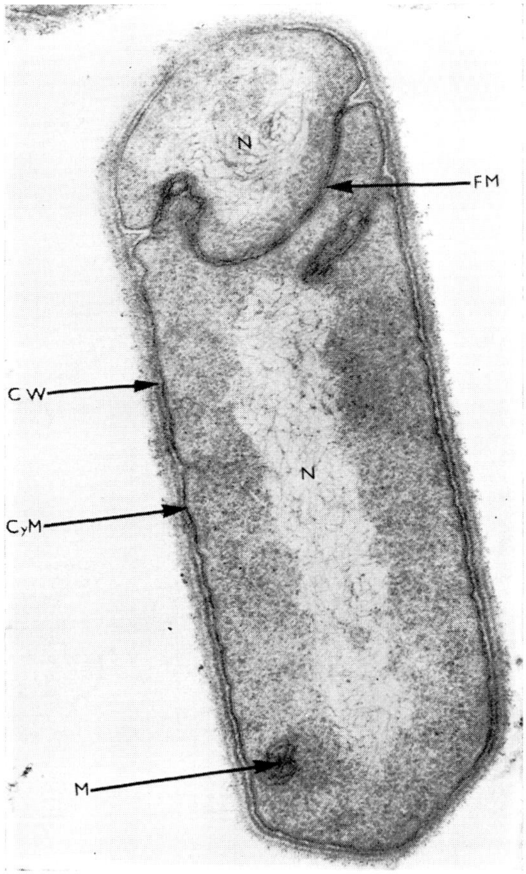

(Reproduced from the Journal of Applied Bacteriology, by permission of Dr P. D. Walker)

FIG. 60
SPORE DEVELOPMENT

Electron micrograph through developing spore and sporangium, slightly more mature than Figure 27. FM, forespore membrane (compare Figure 17); CW, cell wall; CyM, cell membrane; M, mesosome; N, nucleus. The appearance of the longitudinal nuclear rod, in the remainder of the bacillus, is well shown (compare Figures 53 and 54). ×20,000.

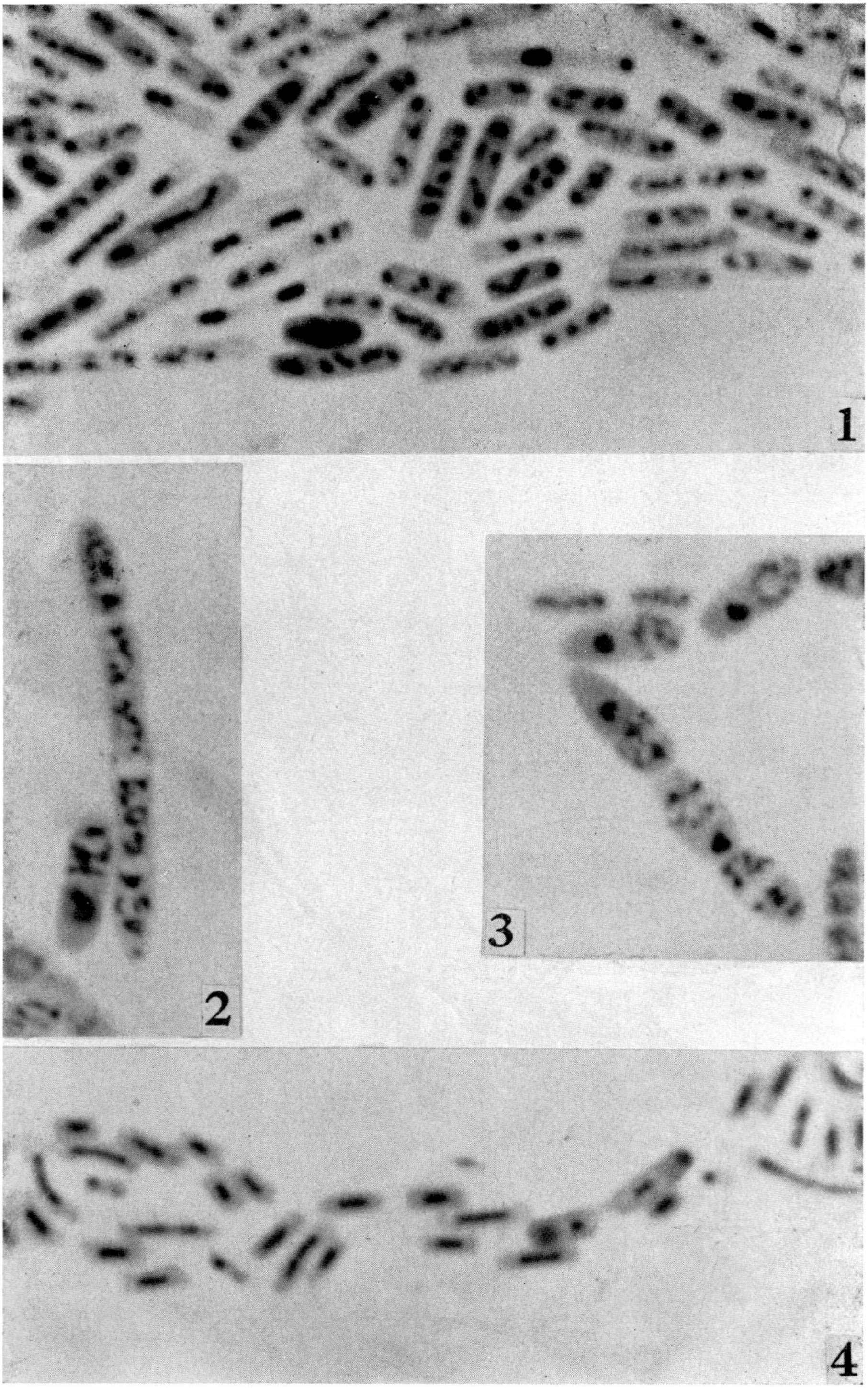

(*Photomicrographs by Dr E. Klieneberger-Nobel, reproduced from the Journal of Hygiene*)

the last section of Chapter 4. As the nuclear thread passes through the replicator (probably with a rotatory action of the whole nucleus around the axis of the bacterium), one of the two daughter threads is thrust, not into a new daughter cell, but through the conjugation bridge into the recipient bacterium. The motive power is the act of DNA replication, at the site of the replicator.(25) Thus, this mode of conjugation can, on this hypothesis, occur only in the actively-growing, vegetative phase of the bacterial cycle, which appears to be the case.

In the last respect, as in several others, the sexual behaviour of *Escherichia coli* seems to be something of an oddity, having little in common with conjugation as it occurs in other protista, or with the various other processes of nuclear fusion that have been observed in bacteria, and which have unfortunately not been accorded the same amount of attention. Rather, it seems to be comparable with two other unusual methods of transfer of genetic material, by virus infection and by 'transformation' (Chap. 10), both of which occur readily in bacteria, and are certainly unusual, if they happen at all, in other living creatures. Probably, the critical factor is simply small size and simplicity of structure. If it is true, as it seems to be (Chap. 4), that the nucleus of *Escherichia coli* is a single molecule of DNA, then it is reasonable to expect that a significant, viable fragment of it might be transferred to another cell, by any one of several simple mechanical actions. In the case of 'transformation', first observed in the pneumococcus, the effect is produced by absorption of a portion of a DNA molecule, derived from a disrupted donor cell. Vegetative conjugation, as described in this section, has more in common with phage action than with any other known sexual process. In both cases, a portion of DNA is injected from one unit into the other, where it parasitises the existing nucleus. Whether this resemblance reflects a true relationship may eventually be elucidated.

THE MATURATION OF THE RESTING NUCLEUS—THE ENDOSPORE

One of the remarkable things about the cycle of changes accompanying the maturation of the resting nucleus in bacteria is that there has been so much agreement between the interpretations of different workers. Most observations have been made upon the bacillary endospore, because it is both obvious and of practical importance, but there is much similarity between it and the various other spores and microcysts that form part of the life-cycles of other groups of bacteria. They seem to resemble and differ from one another in the same way as different microfungi, that have haploid and diploid phases of

---

FIG. 61
THE MATURATION OF THE SPORE

(1) *Clostridium welchii*, vegetative cells, acid-Giemsa ×2800.
(2) *Cl. welchii*, nuclear fusion, acid-Giemsa ×3500.
(3) *Cl. welchii*, fusion cells, acid-Giemsa ×3500.
(4) *Cl. welchii*, maturing sporangia, acid-Giemsa ×1875.

different relative durations [26] (Fig. 39). The resting nucleus appears to be haploid, and germinates to produce a vegetative primary nucleus that is also haploid. In the later stages of culture, there occurs a nuclear fusion resulting (usually, but not invariably) in a longitudinal rod-shaped nucleus, which may be transient or may persist for several generations. Eventually, there is a reduction division and the haploid condition is restored, with the elimination of part of the diploid nucleus. The mature, resting nucleus is then formed [27] (Figs 40 & 43).

In most cases, the fusion process must be regarded as autogamous, since it takes place between the nuclear material of a single cell,

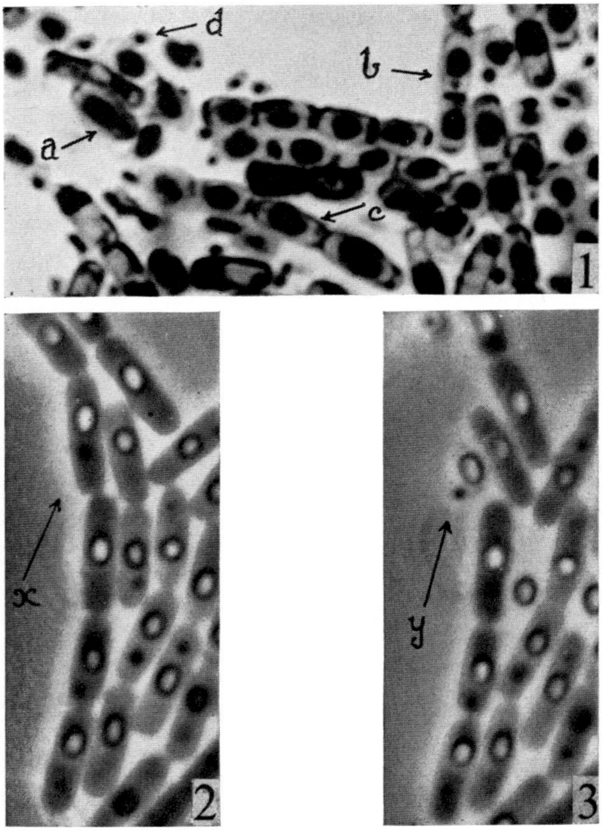

FIG. 62
THE NUCLEAR REDUCTION PROCESS

(1) In *Rhizobium*; *a*, *b*, *c*, *d* represent stages in the maturation of the resting nucleus, and the elimination of a small daughter nucleus. (*Reproduced from Cold Spring Harbor Symposia.*)

(2, 3) The elimination of the rejected daughter nucleus in *Bacillus*; *x*, *y* shows two successive stages in the same organism. Phase-contrast photographs in the living state. (*Reproduced from Experimental Cell Research, by permission of Professor R. J. V. Pulvertaft.*)

but this is not invariably the case. In large, septate bacilli and in some actinomycetes, the nuclei of more than one cell take part in the process, but even these must normally be sister nuclei, recently separated. There is a distinct difference between the septate and unicellular sporing bacilli in this respect. [10]

The early studies of the nuclear process in sporulation, by Stille [28] and Piekarski, [29] did not show the fusion nucleus, although, curiously, the corresponding reduction process was clearly indicated. These workers illustrated a spore with a single, small, central, spherical nucleus, which divided into two, as germination proceeded, and then into four. The bacillus in process of sporulation was shown with two nuclei, of which one became the new spore nucleus, and one was eliminated. This gave a rough outline of the cycle subsequently described by Klieneberger-Nobel, [8] Bisset, [10, 11, 26] Flewett [30] and others, wherein an important and invariable feature was the appearance of a fusion nucleus, in the form of a longitudinal rod, that intervened between the vegetative nucleus and the vesicular resting nucleus of the spore. This fusion nucleus was described as breaking up into two or four, of which one unit was enclosed in the spore, and one or three remained outside and either degenerated or were eliminated (Figs 29, 60 & 61). The reality of this reduction process was confirmed by Pulvertaft, [31] who illustrated it with serial photomicrographs of living material in phase-contrast (Fig. 62). Later work has continued to provide confirmation of the details of this cycle. It is noteworthy that the process of nuclear reduction or elimination was described by DeLamater and Hunter, [32] although their mitotic interpretation of the bacterial nucleus was unorthodox and has been abandoned. The whole cycle was redescribed in exactly the same terms as previously by Young and Fitz-James, [33] who made a most elaborate investigation, combining cytochemical staining, phase-contrast and electron microscopy of ultra-sections: and a more recent study by Ellar and Lundgren, [34] using the latter methods, showed part of the nuclear cycle, including the longitudinal rod-nucleus and the reduction process, in parallel with the development of the spore envelopes, as described in Chapter 3 (Figs 28 & 29).

MICROCYST FORMATION IN MYXOBACTERIA AND EUBACTERIA

Very much less attention has been paid to the process of maturation of the resting cell in non-sporing bacteria, but in this respect the commoner Gram-negative species resemble myxobacteria and cytophagas quite closely. [10, 12, 35] The resting nucleus is contained in a microcyst which may be larger or smaller than the vegetative bacterium. In eubacteria it is usually small, oval or spherical; in some myxobacteria it is oblong, in others spherical. In cytophagas, it is spherical and very large, compared with the vegetative cell (Chap. 7; Figs 63 & 64).

The microcysts of some myxobacteria are contained in a complex fruiting body (Chap. 8), but those of eubacteria and cytophagas are free.

The cytological processes accompanying the maturation of the microcyst are similar in all these cases. The nuclear material forms a chromatinic rod, resembling that which precedes sporulation in Bacillaceae. The rod divides into two spherical, nuclear bodies that again fuse and form the mature, resting nucleus. The process is almost identical in

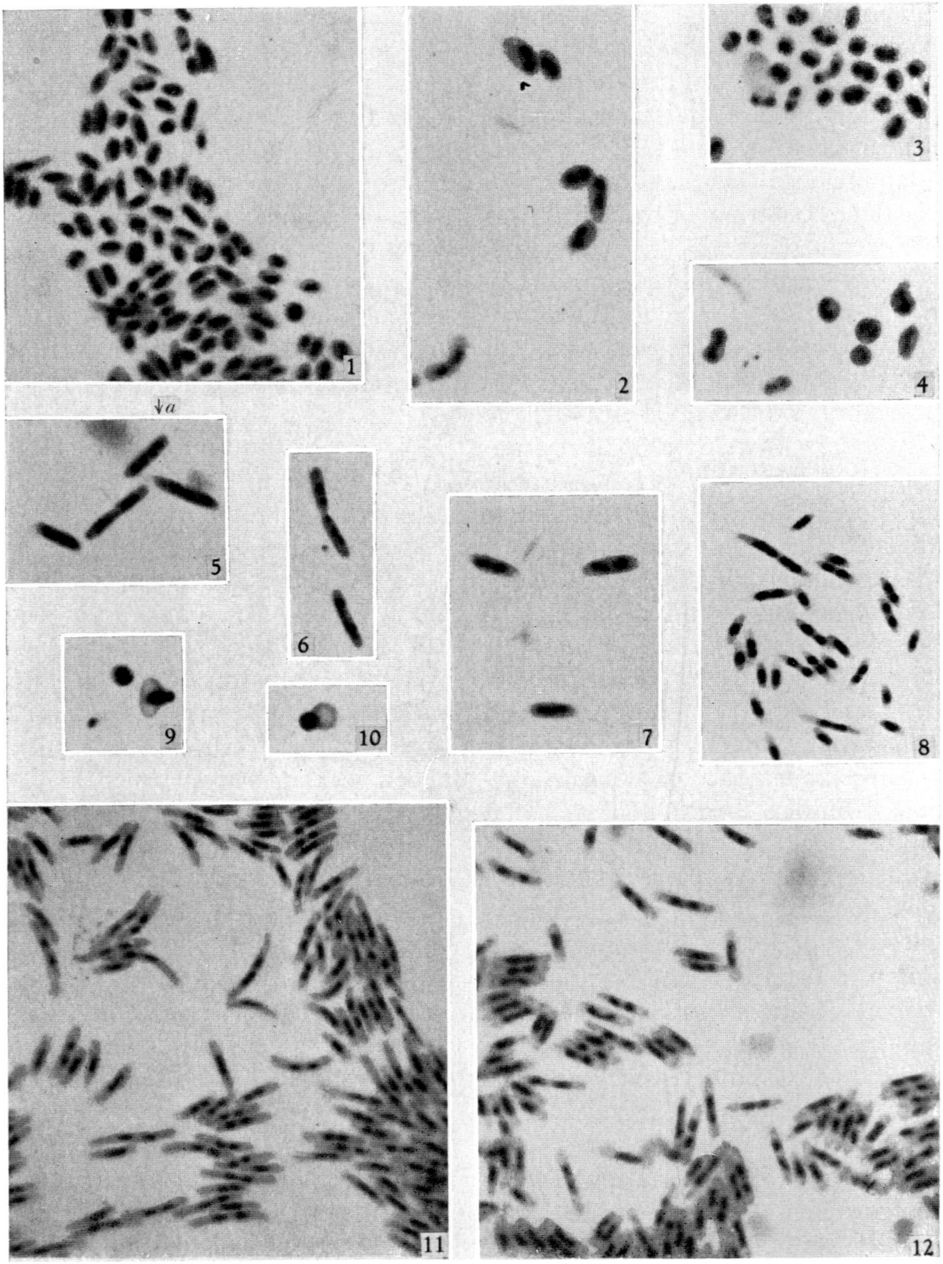

(*Photomicrographs by Dr E. Klieneberger-Nobel, reproduced from the Journal of General Microbiology*)

FIGS. 63, 64

THE CYTOLOGY OF MYXOBACTERIA

(1), (3), (4) *Myxococcus fulvus*, germinating microcysts, Giemsa ×3000.

(2) *Chondrococcus exiguus* (as above).

(5), (6), (8). *M. fulvus*, young vegetative cells, showing chromosomes, Giemsa ×3000.

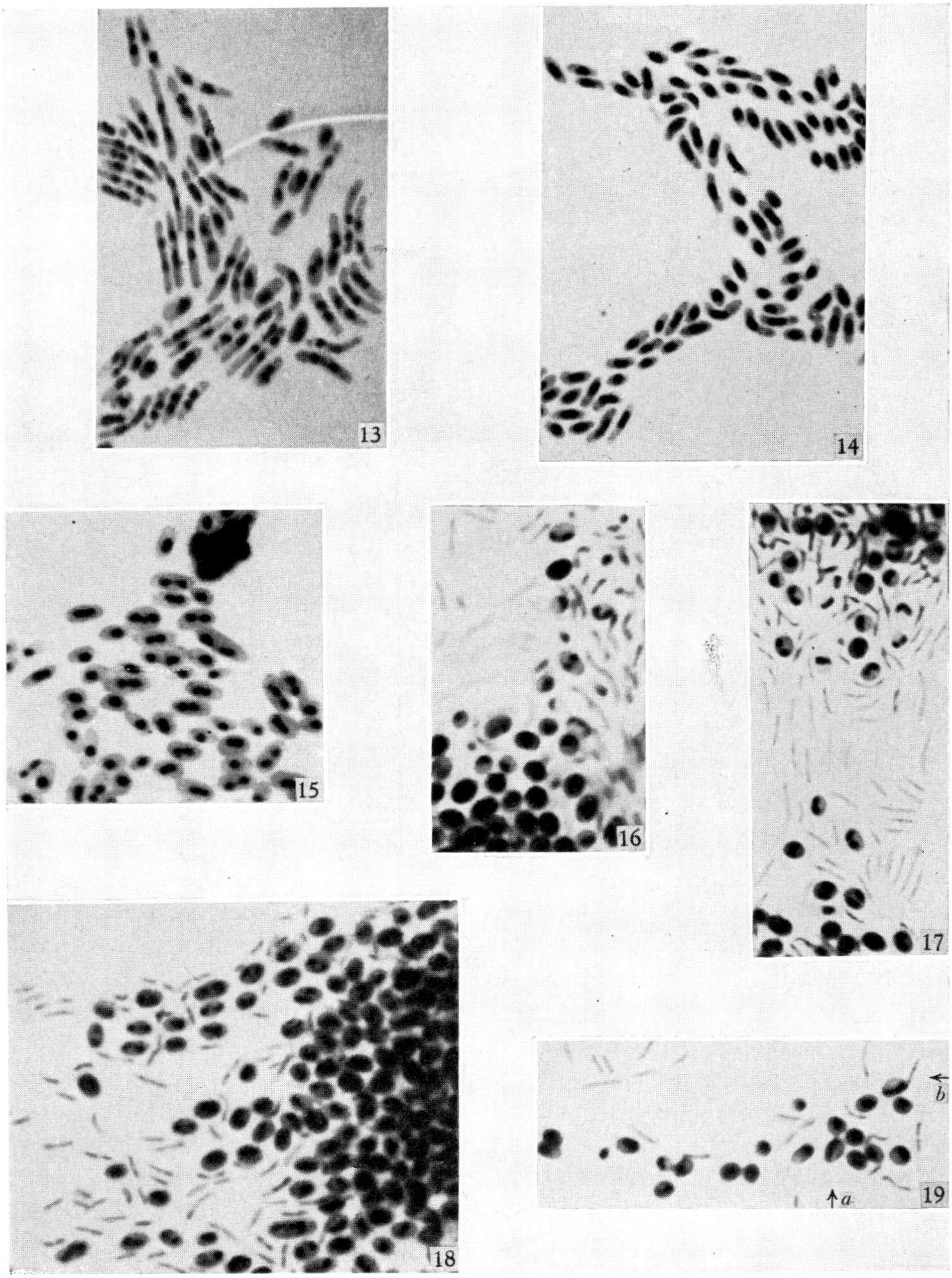

(7) *Ch. exiguus* (as 5).
(9), (10) *M. fulvus*, burst microcysts, Giemsa ×3000.
(11) *M. virescens*, maturing culture, bacilli gathering to form microcysts, Giemsa ×3000.
(12) *M. fulvus*, maturing culture, Giemsa ×3000.
(13), (14), (15) *M. fulvus*, nuclear fusion, Giemsa ×3000.
(16), (17), (18), (19) *M. fulvus*, microcyst formation, Giemsa ×3000.

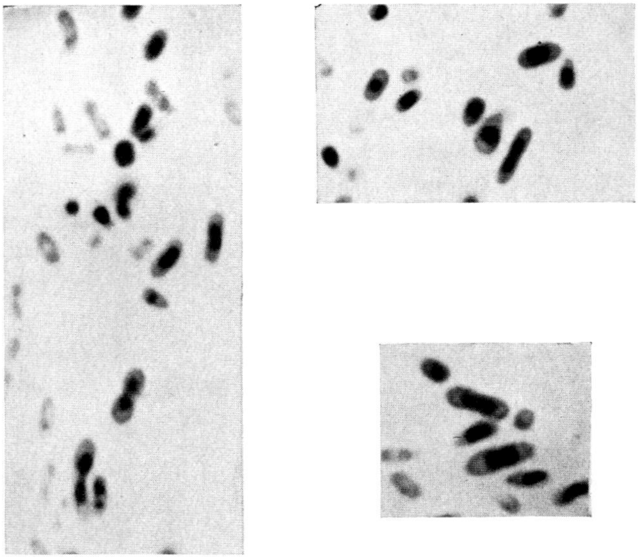

(Reproduced from the Journal of Hygiene

FIG. 65

MATURATION OF THE MICROCYST IN *E. COLI*

Sexual forms. The nuclear material may be seen at various stages of division and refusion. The division is usually incomplete and the fusion autogamous. Occasionally it is complete and the fusion truly sexual.

myxobacteria and most smooth eubacteria, in Nocardia [36] and to some extent in the giant bacterium Oscillospira [37, 38] (Figs 44, 65 & 66).

Two processes of fusion are discernible in the case of the myxobacteria and smooth eubacteria. Firstly, the rod-shaped nucleus arises by the fusion of the nuclear units of the cell, this divides and again fuses. The division of the nucleus may or may not be accompanied by complete division of the cell. If the gametic nuclei fuse within the original cell, then the process is autogamous, if the two gametes separate entirely and conjugate with other partners, then it is sexual.

In eubacteria and cytophagas the division and refusion of the nucleus occurs in an elongated cell, often with a marked, central constriction, but in most myxobacteria the cell is already spherical.

A reduction process almost identical with that found in sporing genera occurs in myxobacteria, cytophagas, non-sporing eubacteria and Nocardia. [27] The nucleus divides into two unequal parts and the smaller of these is eliminated. As in the case of sporers, sometimes more than one body is eliminated.

The resting nucleus may be formed directly from the secondary nucleus, with or without the intervention of a sexual process. In the case of *Bact. malvacearum*, described by Stoughton, [2] the microcyst is extruded from the point of contact of conjugating cells, or from the side of the bacterium when conjugation is not apparent (Fig. 36). As the microcyst grows the mother cells shrink, and eventually disappear. This lateral extrusion of the microcyst has been reported in Bacteriaceae by Mellon [1] but the cytological processes accompanying it were not des-

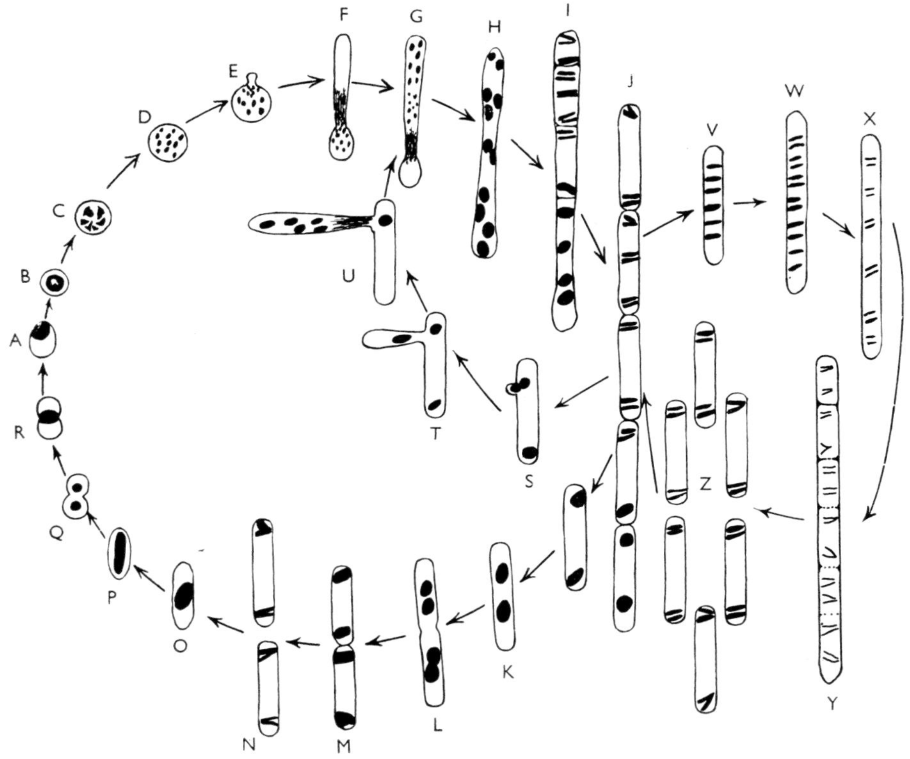

(*Reproduced from the Journal of Hygiene*)

FIG. 66
THE LIFE-CYCLE OF *NOCARDIA*

A. Microcyst.
B-G. Germination.
G-J and V-Z. Simple and complex fission.
S-U. Branching.
K-N. Simple fission in later vegetative stages.
O-A. Maturation of the microcyst.
In most respects *Nocardia* behaves in a similar manner to the eubacteria and myxobacteria, but branching and germination by germ tube are typical of higher bacteria. (According to Morris.)

cribed, because, at that time, there existed no reliable techniques for the demonstration of the bacterial nucleus.

NUCLEAR FUSIONS IN VEGETATIVE REPRODUCTION

In the previous chapter, an apparently syngamous method of vegetative division was described. In an elongated cell, six nucleoids fuse, separate, divide, and are redistributed to six daughter bacilli that arise by fragmentation of the filament. This appears to be merely an alternative method of reproduction to simple fission, and is analogous to the formation and dispersion of a symplasm, although the characteristic form of the bacterial nucleus and the diameter, if not the length of the cell is preserved throughout the process (Figs 47-50).

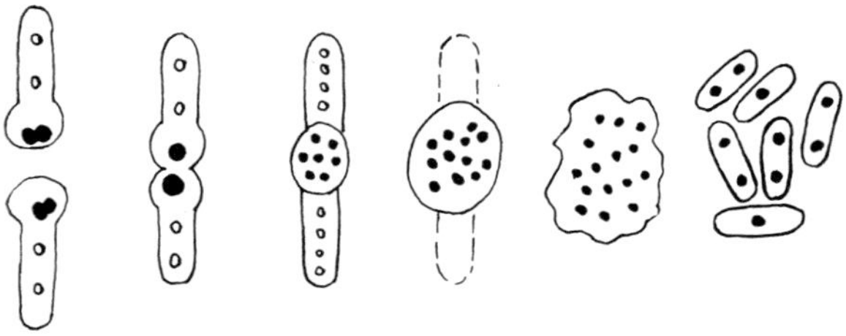

FIG. 67
CONJUGATION IN *PROTEUS*

Sexual conjugation in the vegetative phase of *Proteus*, after Rolly.

A comparable, but entirely distinct method of vegetative reproduction, with the intervention of a symplasm, preceded by an apparently sexual fusion, has been quite frequently reported to occur in Gram-negative bacteria, and has been fully illustrated by independent research workers.[39, 40, 41] The most complete description is that of Rolly;[39] the initial stage is the production of a large swelling at the pole of a bacterium. This contains a single enlarged nucleus. Two such bacteria become fused together at the swollen poles, and their nuclei apparently conjugate. The result is a large sac-like body, with the residual part of the bodies of two bacteria protruding from it. This sac continues to enlarge and becomes irregular in form. The nucleus divides several times, and eventually the symplasm gives rise to a new generation of several bacteria (Fig. 67).

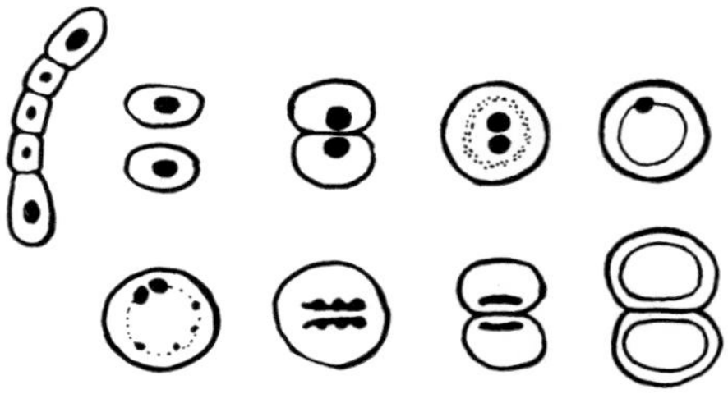

FIG. 68
MATURATION OF THE RESTING CELL IN *M. TUBERCULOSIS*

Upper line—Sexual conjugation of cells derived from the bacillary form; formation of the vesicular nucleus.

Lower line—Reduction division. Two large chromosomes are formed from the vesicular nucleus and one passes to each daughter cell. The vesicular nucleus is reconstituted in each; lacking the typical eccentric granule. Modified after Lindegren and Mellon.

This process entails a fusion between only two bacteria, but it strongly resembles the so-called star-formation that occurs commonly in pseudomonads, especially plant-pathogens,[3-7, 42] and has been claimed to be a sexual process, although it may entail polar fusion between several bacteria simultaneously (Fig. 57). Braun and Elrod[42] have shown that the Feulgen-positive material con-

REFERENCES

1. MELLON, R. R. (1925). *J. Bact.* **10**, 579.
2. STOUGHTON, R. H. (1932). *Proc. R. Soc.* **B**, **111**, 46.
3. BEIJERINCK, M. W. (1890). *Bot. Ztg*, **48**, 837.
4. STAPP, C. (1942). *Zentbl. Bakt. ParasitKde, II*, **105**, 1.
5. STAPP, C. & KNÖSEL, D. (1954). *Zentbl. Bakt. ParasitKde, II*, **108**, 243.
6. HEUMANN, W. (1956). *Arch. Mikrobiol.* **24**, 362.
7. HEUMANN, W. (1962). *Z. VererbLehre*, **93**, 441.
8. KLIENEBERGER-NOBEL, E. (1945). *J. Hyg., Camb.* **44**, 99.
9. KLIENEBERGER-NOBEL, E. (1947). *J. gen. Microbiol.* **1**, 33.
10. BISSET, K. A. (1949). *J. Hyg., Camb.* **47**, 182.
11. BISSET, K. A. (1950). *J. gen. Microbiol.* **4**, 1.
12. GRACE, J. B. (1951). *J. gen. Microbiol.* **5**, 519.
13. LEDERBERG, J. & TATUM, E. L. (1946). *Cold Spring Harb. Symp. quant. Biol.* **11**, 113.
14. LEDERBERG, J. (1956). *J. Bact.* **71**, 497.
15. ANDERSON, T. F., WOLLMAN, E. L. & JACOB, F. (1957). *Annls Inst. Pasteur, Paris*, **93**, 450.
16. HAYES, W. (1952). *Nature, Lond.* **169**, 118.
17. WOLLMAN, E. L. & JACOB, F. (1957). *Annls Inst. Pasteur, Paris*, **93**, 323.
18. LEDERBERG, J. (1957). *Proc. nat. Acad. Sci. U.S.A.* **43**, 1060.
19. HAYES, W. (1960). *Symp. Soc. gen. Microbiol.* **10**, 12.
20. HAYES, W. (1953). *J. gen. Microbiol.* **8**, 72.
21. WOLLMAN, E. L., JACOB, F. & HAYES, W. (1956). *Cold Spring Harb. Symp. quant. Biol.* **21**, 141.
22. JACOB, F. & WOLLMAN, E. L. (1958). *Symp. Soc. exp. Biol.* **12**, 75.
23. JACOB, F. & BRENNER, S. (1963). *C. r. hebd. Séanc. Acad. Sci. Paris*, **256**, 298.
24. JACOB, F., BRENNER, S. & CUZIN, F. (1963). *Cold Spring Harb. Symp. quant. Biol.* **28**, 329.
25. GROSS, J. D. & CARO, L. (1965). *Science, N.Y.* **150**, 1679.
26. BISSET, K. A. (1951). *Cold Spring Harb. Symp. quant. Biol.* **16**, 373.
27. BISSET, K. A., GRACE, J. B. & MORRIS, E. O. (1951). *Expl Cell Res.* **2**, 388.
28. STILLE, B. (1937). *Arch. Mikrobiol.* **8**, 125.
29. PIEKARSKI, G. (1940). *Arch. Mikrobiol.* **11**, 406.
30. FLEWETT, T. H. (1948). *J. gen. Microbiol.* **2**, 325.
31. PULVERTAFT, R. J. V. (1950). *J. gen. Microbiol.* **4**, 14.
32. DELAMATER, E. D. & HUNTER, M. E. (1952). *J. Bact.* **63**, 13.
33. YOUNG, I. E. & FITZ-JAMES, P. C. (1959). *J. Biophys. Biochem. Cytol.* **6**, 467.
34. ELLAR, D. J. & LUNDGREN, D. G. (1966). *J. Bact.* **92**, 1748.
35. KLIENEBERGER-NOBEL, E. (1947). *J. gen. Microbiol.* **1**, 1.
36. MORRIS, E. O. (1951). *J. Hyg., Camb.* **49**, 175.
37. DELAPORTE, B. (1934). *C. r. hebd. Séanc. Acad. Sci. Paris*, **198**, 1187.
38. TUFFERY, A. A. (1954). *J. gen. Microbiol.* **10**, 342.
39. ROLLY, H. (1952). *Zentbl. Bakt. ParasitKde I, Orig.* **157**, 586.
40. SMITH, W. E. (1944). *J. Bact.* **47**, 417.
41. HUTCHINSON, W. G. & STEMPEN, H. (1954). *Symp. Am. Ass. adv. Sci.* **29**.
42. BRAUN, A. C. & ELROD, R. P. (1946). *J. Bact.* **52**, 695.
43. LINDEGREN, C. C. & MELLON, R. R. (1932). *J. Bact.* **25**, 47.
44. LINDEGREN, C. C. & MELLON, R. R. (1933). *Proc. Soc. exp. Biol. Med.* **30**, 110.
45. NEDELKOVITCH, J. (1950). *Annls Inst. Pasteur, Paris*, **78**, 177.
46. BREED, R. S., MURRAY, E. G. D. & SMITH, N. R. (1957). *Bergey's Manual of Determinative Bacteriology*, 7th ed. London: Baillière, Tyndall & Cox.
47. ALEXANDER-JACKSON, E. (1954). *Growth*, **18**, 37.
48. MORRIS, E. O. (1951). *J. Hyg., Camb.* **49**, 46.
49. BATTY, I. (1958), *J. Path. Bact.* **75**, 455.
50. PRÉVOT, A. R. (1953). Symposium. Actinomycetales, *6th Int. Congr. Microbiol.*
51. KLIENEBERGER-NOBEL, E. (1947). *J. gen. Microbiol.* **1**, 33.
52. DICKENSON, P. B. & MACDONALD, K. D. (1955). *J. gen. Microbiol.* **13**, 84.
53. BISSET, K. A. (1957). *J. gen. Microbiol.* **17**, 562.
54. DIENES, L. (1946). *Cold Spring Harbor Symp. quant. Biol.* **11**, 51.
55. KVITTINGEN, J. (1949). *Acta Path. Belgr.* **26**, 24 & 855.
56. MORRIS, E. O. (1953). *J. Hyg., Camb.* **51**, 49.
57. DIENES, L. & BULLIVANT, S. (1967). *Ann. N.Y. Acad. Sci.* **143**, 719.

CHAPTER 6

Reproduction

THE GROWTH CYCLE

The growth cycle of bacteria, whether in culture or under natural conditions, follows a regular pattern of which the main outlines have long been known, although the underlying chemical and cytological changes have more recently been discovered. [1-5]

When bacteria are transplanted upon a new medium, suitable for their growth, from an older culture which has passed its period of active reproduction, there is at first an interval of time, known as the lag phase, in which no numerical increase occurs. The bacteria may increase in size but do not divide. The lag phase lasts from one or two hours to six or seven or longer, after which the logarithmic phase commences. The bacteria reproduce by fission, sometimes at very short intervals, and increase in numbers at an approximately logarithmic rate. Much later, after a period which is a function of food-supply, temperature and the physical conditions in the culture, the rate of growth falls steadily and eventually almost ceases. In the decline phase the numbers may actually decrease, although they usually remain static for a long period.

It has been discovered that the chemical constitution of the cells, and especially their nucleotide content varies according to a similar cycle. Bacteria in aged cultures have a low nucleotide content, which, when they are transplanted to a new medium, rises to a high level during the course of the lag phase and falls off gradually as the culture ages. The nucleotide content thus corresponds closely to the state of activity of the nucleus, and presumably indicates the actual level of nuclear material in the cell.

As already indicated (Chap. 4), the lag phase represents the period of germination of the microcyst. The metabolic activity and nucleotide content of the resting nucleus are both low, but in the young, vegetative cell both are high.

If the medium is inoculated, not with cells from an aged culture, but with those already in the active condition, the lag does not occur. The nuclear material is already in the reproductive phase and no delay is entailed.

SIMPLE VEGETATIVE REPRODUCTION

The enormously rapid increase in numbers, which occurs in the logarithmic phase of a bacterial culture, attests to the efficiency of simple fission as a method of reproduction. This rapidity of growth is, of course, due mainly to the small size of bacteria, and the consequent high ratio of cell surface to volume. Rapid colonisation of a new medium is, nevertheless, assisted by the means of reproduction employed.

The epithet ‘simple’, although habitually employed to describe reproduction by fission, is less accurate than once it was believed to be. Many bacteria are multicellular. [6-9] Corynebacteria and mycobacteria are composed of from one to a dozen small cells; eubacteria of rough morphology are normally four-celled, and even cocci may contain two, three or four cells. In these circumstances mere cell division does not provide either increased surface area or wider distribution in the medium unless accompanied by fission of the bacterium. In Gram-negative bacteria multicellularity is mainly associated with active reproduction in very young cultures [1, 2] and is seldom evident in older cultures when it might be expected to be disadvantageous, under conditions of more intense competition for a diminished supply of nutrients.

Division can never be truly simple, but must always entail division of the nucleus, the formation of a transverse, membranous septum and the secretion of new cell walls, at the same time as the growth of the cell must itself continue. Complementary to the production of structures peculiar to cell division the entire organism increases in size, extends its membrane and wall, its cytoplasm and nuclear material. It has already been explained in Chapter 3 that bacterial cell division is not always equational, and may resemble a budding process.

Bacteria possess a marked polarity, and almost invariably divide transversely. This is also true of many cocci, which elongate and divide always in the same plane, although in others, notably the commoner types of staphylococci, each division is at right-angles to the previous one. [6, 8] This results in the production of the typical, grape-like clusters. At least one filamentous bacterium, Dermatophilus, also divides in the same manner as a staphylococcus. [9] Rod-shaped bacteria, growing under adverse conditions, may produce pathological forms which divide in an irregular manner, but this rarely occurs in healthy cultures.

Asexual reproduction by fission is common to most unicellular or simple organisms, and is frequently found to alternate with a sexual process, as it does in bacteria also. Multicellular plants and animals have short-circuited this cycle, to some extent, by the production of specialised sexual cells, upon which alone falls the duty of perpetuating the species. The somatic cells reproduce vegetatively and asexually, but eventually die, whereas the reproductive cells, after a brief vegetative career, may undergo sexual conjugation and survive. In bacteria, which lack this specialisation, all cells alike possess the potential for both vegetative and sexual reproduction, although, as in most other cases, few of the cells which achieve maturity in the resting stage are likely to be transferred from a declining culture to a fresh medium, and serve to initiate a new, vegetative generation.

POST-FISSION MOVEMENTS

Attempts by several workers to study the mechanism of cell division in living bacteria have caused an undue amount of attention to be directed towards an interesting, but quite unimportant artefact, the so-called post-fission movements. The phenomenon was described in 1910, by Graham-Smith [10] who perfectly understood its artificial nature. It has, however, been re-examined by workers who appear to have misinterpreted its significance entirely, and it was at one time accorded undue prominence in many elementary textbooks, probably in default of more valuable information upon cell division in bacteria.

It is stated that, after division, the daughter bacteria move, with relation to one another, in one of several different and characteristic ways. Smooth forms slip past one another and come to lie side by side; rough bacilli

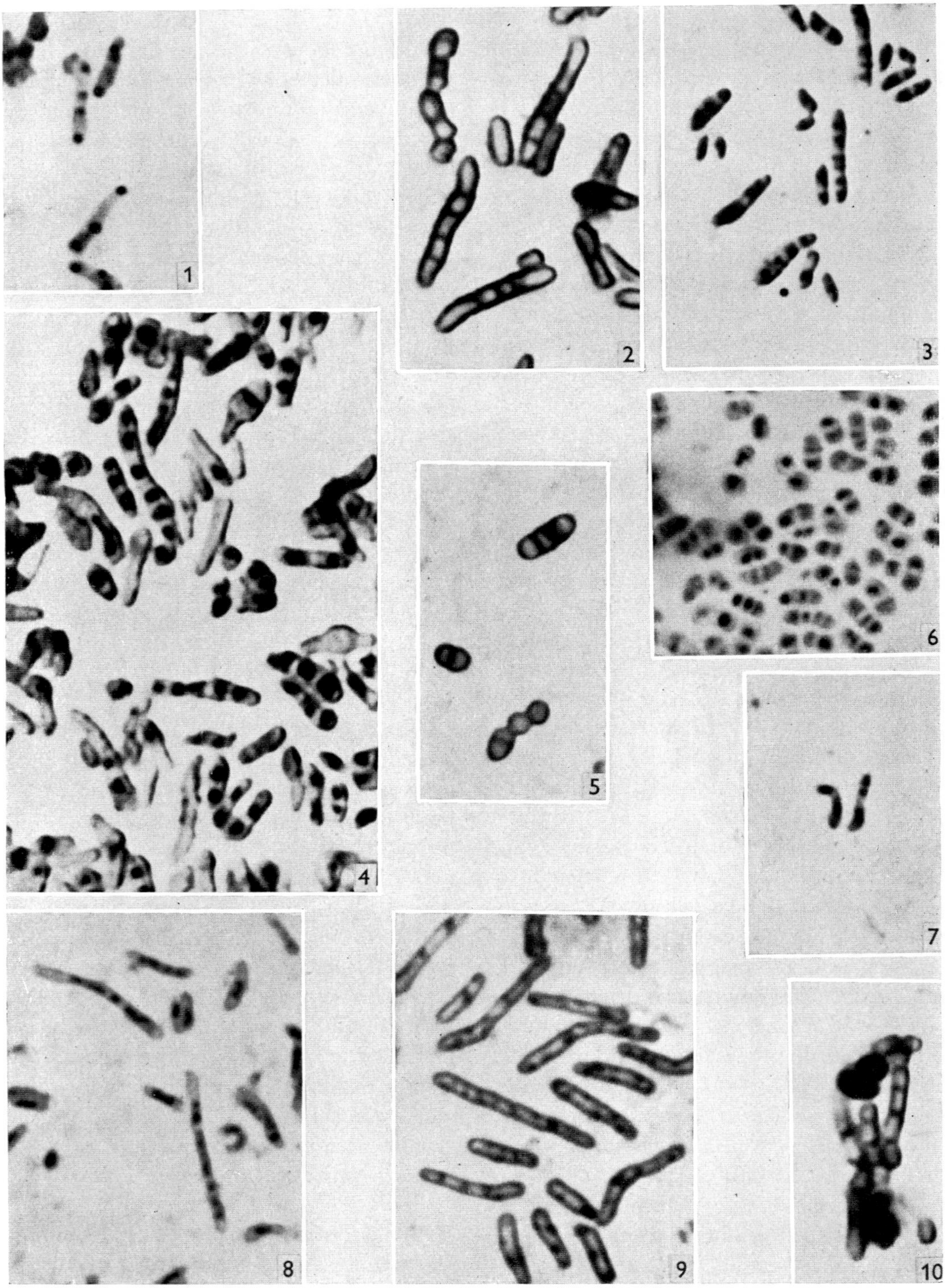

(Reproduced from the Journal of General Microbiology)

move as upon a hinge at the point of division, the 'snapping' movement; and others, notably corynebacteria, perform this second movement in an exaggerated form, so that the two halves move round upon one another like the closing of a pocket-knife.

These movements may be observed quite readily, but only under the correct conditions. If bacteria are examined in fluid culture or upon the surface of an agar plate, nothing of the kind will be seen. But if the same bacteria are examined either when set in the thickness of solid medium, or growing between a block of medium and a coverslip upon its surface, then the post-fission movements will become obvious. It was by the employment of these techniques that they were first observed, and while it is entirely accurate to describe bacteria as behaving in this manner, under these conditions, the assumption that they do so under normal cultural conditions is quite unjustifiable. The post-fission movements are merely the result of the growth of the bacteria under conditions of severe mechanical restraint. Smooth bacteria grow and elongate against the pressure of the surrounding medium. When separation is complete, the daughter cells may be forced back again, side by side, by the elasticity of the agar. Rough bacilli remain attached at the point of division, and the filament, constrained in the same manner, is compressed concertina-fashion, bending at the points of division. The division of corynebacteria is more complex and will be discussed in the section on Reproduction in Septate Bacteria, in this chapter. It is sufficient here to say that the centre of the bacillus may be very flexible, by reason of its multicellular structure, and the two halves thus remain attached even when forced into an acute angle.

The mechanical constraint which brings about these appearances, by its opposition to the growth of the bacteria, occurs to a very limited degree in growth upon the surface of solid medium, and causes those pale shadows of post-fission movements which contribute to the architecture of bacterial colonies [11, 12] (Chap. 8 and Fig. 89).

COMPLEX VEGETATIVE REPRODUCTION

One of the most striking of the complex reproductive processes that accompany and supersede simple fission in the later stages of a bacterial culture is the filamentous growth and fragmentation, [2, 13, 14, 15] to which reference has already been made (Chap. 4) and which occurs in many different types of bacteria, in one of a number of similar forms.

FIG. 72
THE CYTOLOGY OF CORYNEBACTERIA AND MYCOBACTERIA

(1) *C. diphtheriae* stained by Neisser's stain, after heat fixation. The 'typical morphology' and 'metachromatic granules' are seen.

(2), (3), (4) *C. diphtheriae* stained cytologically for cell walls (by tannic-acid-violet (2)) and for the nuclear structures (by acid-Giemsa, (3), (4)). The bacilli are multicellular, and reproduce in two characteristic ways; by cell proliferation and simple fission, or by fragmentation.

(5), (6) Similar preparations of a vaginal corynebacterium.

(7), (8), (9) Similar preparations of *M. tuberculosis*. Its structure resembles that of the corynebacteria.

(10) Cell wall stain of *M. phlei*.

All at ×3000.

Fusion cells appear, containing three pairs of chromosome complexes, in the case of smooth eubacteria, and similar fusion nuclei in other types of bacteria. The fusion cell grows into a long filament, undergoing two nuclear divisions, and eventually fragments into individual bacteria, usually in multiples of three.

As this process includes an appearance of nuclear fusion and reorganisation it may be regarded as sexual reproduction. It is, however, a vegetative process, directed towards the purpose of increasing the number of bacteria in the culture, and not as in the case of the later sexual fusions, concerned with the distribution and perpetuation of the species. As a method of reproduction it probably is not much less efficient than simple fission, because it does not hinder growth by any serious reduction in the proportion of cell surface to volume, as the formation of a more typical symplasm might do. Some energy must be required for the initial fusion, but this need not necessarily be a great loss, and the disadvantage is presumably outweighed by the advantage of the redistribution of nuclear material.

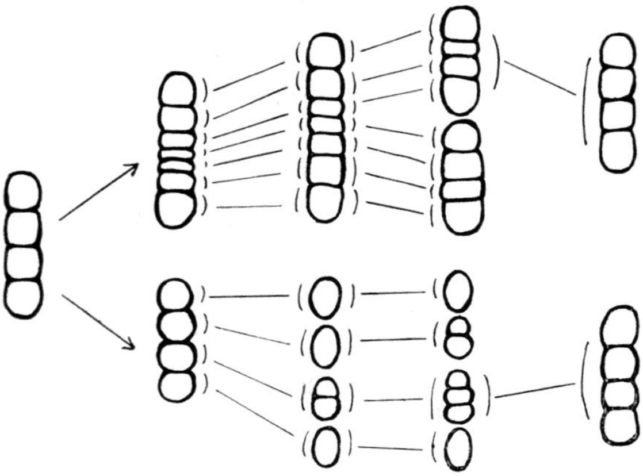

(*Reproduced from the Journal of General Microbiology*)

FIG. 73
ALTERNATIVE MODES OF DIVISION IN MYCOBACTERIA AND CORYNEBACTERIA

Top—Proliferation of cells, followed by simple fission. *Below*—Fragmentation into individual cells, which by growth and division return to the original condition.

As the behaviour of the chromosome complexes during simple fission is such as to suggest that a rapid segregation of heterozygotes would probably occur, it is possible that sexual, vegetative reproduction may assist in the redistribution of genetical characters in the culture, prior to the change of nuclear phase which usually follows.

This mode of reproduction may be found at any stage of culture, but is most common in cultures aged from 12 to 48 hours, and may often supersede simple fission almost entirely. Cultures in this condition develop a macroscopically rough appearance, because of the high proportion of filamentous cells; but this is transient, and disappears as the culture matures.

In myxobacteria, filamentous cells bearing this type of nuclear structure move actively with the rest of the swarm, and exhibit sufficient co-ordination to enable them to crawl as a unit. There is much more apparent variation in the form of such fusion cells in myxobacteria than in eubacteria, and they are rather less frequent.

In *Sphaerotilus natans*, a chlamydobacterium, the form of the fusion cells and their nuclei is similar to that of smooth eubacteria, but their behaviour is not known.

REPRODUCTION IN SEPTATE BACTERIA

Mycobacteria, corynebacteria and nocardias possess a morphology which differs strikingly from that of other rod-shaped bacteria. Bacilli of these genera may consist of a single, spherical or oval cell, or of a dozen or more such cells. [6, 16, 17] Often the terminal cells are much larger than the others, and the contents of such enlarged, terminal cells constitute the metachromatic granules of *C. diphtheriae*. In the case of *M. tuberculosis* the cells are more usually similar in size, or approximately so, and the beaded appearance of the bacillus in a preparation stained by Ziehl-Neelsen's method is due to shrinkage of the cell contents (Figs 1 & 72).

Vegetative reproduction in corynebacteria may take one of two different forms, that correspond to the simple and complex types of fission in eubacteria, although the resemblance is not exact. [16]

In actively growing cultures there is rapid cell division at the centres of the bacilli, where the cells become very numerous, without much growth of the bacterium, so that some cells are reduced to narrow discs, although the terminal cells retain their spherical or oval shape. This multiple cell division is followed by fission of the bacterium. The new terminal cells, initially small, increase in size, while the central cells continue to multiply rapidly, and the process is repeated. This corresponds to simple fission in unicellular bacteria, and the two processes are not readily distinguishable by routine bacteriological methods (Fig. 73).

In mycobacteria and nocardias the difference in size, between the central and terminal cells is not so great, but the mode of division is otherwise similar. [16, 17, 18] Several authors have recorded that tubercle bacilli in culture alternate between shorter and longer bacilli, but interpretations of the significance of this phenomenon differ very greatly. [19, 20]

The alternative method of vegetative reproduction may be found simultaneously, in

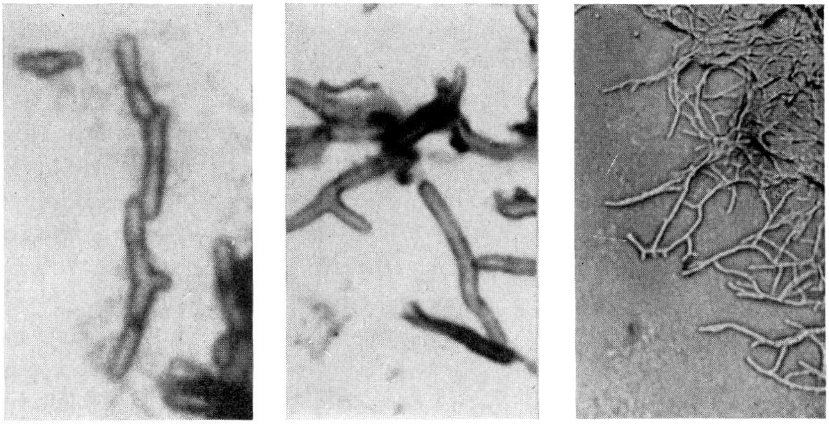

A B C

(B and C reproduced from the Journal of General Microbiology)

FIG. 74
BRANCHING IN BACTERIA

A. Branching of the transient, budding type in *C. diphtheriae*. Tannic-acid-violet ×3000.
B. Somewhat more advanced type of transient branching in a pathogenic actinomyces. Here a true branch may form, but rapidly becomes detached from the main filament. Tannic-acid-violet ×3000.
C. Permanent branching in a streptomyces. Unstained, *in situ* ×1000.

the same culture, although different strains of bacteria show a preference for one method or the other. In this case the individual cells grow until they are all spherical or oval in shape. They then separate completely from one another and develop into typical, multicellular bacilli. This is superficially comparable with sexual, vegetative reproduction in eubacteria, but the analogy is by no means complete. There is no apparent sexual process involved, and the filament which fragments to produce the new bacilli is not unicellular, but is a chain of cells, differing only in their larger size from those composing the original bacillus from which the filament arose. It appears, in fact, to be a vegetative process as simple as, or simpler than the first (Fig. 73).

There is general agreement that nocardias have a resting stage, that is achieved after a phase of growth in non-septate coenocytic filaments, [17, 21] but this part of the life-cycle of the tubercle bacillus, like most other things about it, remains controversial. Wyckoff and Smithburn [22] claimed that the resting stage of *M. phlei* arises from the fragmentation of the bacilli, and consists of a small, coccal body. This is confirmed by several other workers [23, 24] and it appears probable that the coccal body has been described as a Gram-positive coccus upon occasion (Chap. 5). The problem of life-cycles in mycobacteria will be discussed also in Chapter 7.

BUDDING AND BRANCHING

There is a certain amount of evidence, discussed at length in Chapter 3, to suggest that the division of bacteria is seldom entirely equational, but that more new material, in the form of cytoplasm, [25, 26] flagella [27, 28] and possibly cell envelopes, [27, 29] goes with one daughter bacterium than with the other. Thus the so-called simple fission of bacteria has certain characters in common with the budding of yeasts. Branching can also occur under adverse conditions in eubacteria [30] but both branching and a variety of budding that is associated with it are quite common in filamentous bacteria [20] and can even be found in rough variants of spirilla [31] although true, permanent branching is uncommon except in Actinomycetales. Some corynebacteria branch in a fashion (Fig. 74) and the so-called 'Lactobacillus bifidus' may be covered with buds and branches; [32] in contrast with true lactobacilli, that are usually septate and often filamentous, but seldom or never branched. [33] These branches are not permanent.

In the anaerobic Actinomycetes [34] the branch appears first as a small, lateral bud in the cell wall. It elongates, but before it attains to any great size, a septum is formed, dividing it from the parent cell. The branch eventually breaks off at this point and continues to grow as a separate filament. This method of branching is quite unlike that of streptomyces or fungi, in which the mature branch remains attached to the main stem, and in which the cell itself is branched. It more closely resembles the budding of yeasts, and results in the production, not of a stable mycelium, but of an increased number of individual filaments. It may be considered to be a device designed to permit simple, vegetative fission in a filamentous type of bacterium.

The same type of branching is also found in 'soil diphtheroids', so-called. These bacteria have little resemblance to *C. diphtheriae* but have this, and other characters in common with actinomyces.

Branching in streptomyces is a permanent and integral part of their structure, and it results in the formation of a fungus-like mycelium, which may be regarded as a single organism. [35] Cell division takes place by the production of transverse septa, often at considerable intervals, occasionally close together.

Branching in mycobacteria and corynebacteria is usually even more transient than in the case of actinomyces, and is confined to a number of specialised strains. It is obvious that microscopic appearances suggestive of branching must be regarded with caution when they occur in bacterial genera that possess so complex a structure, and especially when this structure is not made apparent by standard techniques. Many of the reports of branching in these genera are thus of doubtful value. One of the main exceptions is the avian strain of *M. tuberculosis* described by Brieger and Fell, [18] which branches freely. In this case the appearance suggests that the branch is usually divided from the main stem by a cell wall. Branching is unusual, although not unknown in human strains of *M. tuberculosis*, but seldom results in the formation of structures more complex than a Y-shaped bacillus. A 'branching' strain of *C. diphtheriae*, examined by the author, appeared perfectly normal and unbranched except in very young cultures, where, by ordinary staining methods, a picture was produced, suggestive of extremely profuse branching. When stained by cell-wall stains, however, the appearance was of masses of adherent bacilli, and occasional, small, lateral excrescences or buds in the cell wall. These buds developed, most frequently, in positions where the normal growth of the cells was obstructed by the adhesion of neighbouring bacteria. They were probably no more than outgrowths of cells unable to increase in size by elongation, in the ordinary manner. A variation of this curious type of branching is found in *Leptotrichia dentium*. [36] The septate filament becomes zig-zag, and buds appear at the elbows. These grow into narrow filaments, and the whole complex breaks up into bacilli, each with a thick and a thin portion.

It is of some importance that the definition of branching, as a taxonomic character, should under no circumstances be based upon heat-fixed, Gram-stained material, nor, preferably, upon the evidence of smears. Whenever possible, whole mounts of colonies should be used, stained by an appropriate cytological technique. The very frequent use of the epithet 'tangled', in qualification of the descriptive term 'mycelium' is an adequate commentary upon the disinclination of research workers to differentiate between the natural effects which they wish to study, and the artefacts which inevitably arise from the mishandling of biological material.

SUMMARY

The lag phase of the growth cycle corresponds to the period of germination of the microcyst, the logarithmic phase to the period of nuclear activity, and in the decline phase the resting condition of the nucleus is adopted. The nucleotide content of the cell is correlated with these changes in nuclear activity.

Vegetative reproduction may be asexual or sexual, simple or complex. Analogous methods of fragmentation and regeneration occur in simple eubacteria and in septate bacteria. Post-fission movements are artefacts.

Streptomyces form a complex, branched mycelium, but branching in most other groups of bacteria is usually impermanent and analogous to budding.

REFERENCES

1. ROBINOW, C. F. (1942). *Proc. R. Soc.* **B**, **130**, 299.
2. ROBINOW, C. F. (1944). *J. Hyg., Camb*, 43, 413.
3. MALMGREN, B. & HEDEN, C. G. (1947). *Acta path. microbiol. scand.* **24**, 437.
4. MALMGREN, B. & HEDEN, C. G. (1947). *Nature, Lond.* **159**, 577.
5. BISSET, K. A. (1949). *J. Hyg., Camb.* **47**, 182.
6. BISSET, K. A. & HALE, C. M. F. (1953). *Expl Cell Res.* **5**, 449.
7. WEBB, R. B. (1953). *J. Bact.* **67**, 252.
8. JÄRVI, O. & LEVANTO, A. (1950). *Acta path microbiol. scand.* **27**, 31.
9. RICHARD, J. L., RITCHIE. A. E. & PIER, A. C. (1967). *J. gen. Microbiol.* **49**, 23.
10. GRAHAM-SMITH, G. S. (1910). *Parasitology*, **3**, 17.
11. BISSET, K. A. (1938). *J. Path. Bact.* **47**, 223.
12. BISSET, K. A. (1939). *J. Path. Bact.* **48**, 427.
13. BISSET, K. A. (1948). *J. Hyg., Camb.* **46**, 173.
14. BISSET, K. A. (1948). *J. Hyg., Camb.* **46**, 264.
15. KLIENEBERGER-NOBEL, E. (1945). *J. Hyg., Camb.* **44**, 99.
16. BISSET, K. A. (1949). *J. gen. Microbiol.* **3**, 93.
17. MORRIS, E. O. (1951). *J. Hyg., Camb.* **49**, 175.
18. BRIEGER, E. M. & FELL, H. (1945). *J. Hyg., Camb.* **44**, 158.
19. BRIEGER, E. M., COSSLETT, V. E. & GLAUERT, A. M. (1954). *J. gen. Microbiol.* **10**, 294.
20. LACK, C. H. & TANNER, F. (1953). *J. gen. Microbiol.* **8**, 18.
21. WEBB, R. B. & CLARK, J. B. (1957). *J. Bact.* **74**, 31.
22. WYCKOFF, R. W. G. & SMITHBURN, K. C. (1933). *J. infect. Dis.* **53**, 201.
23. NEDELKOVITCH, J. (1950). *Annls Inst. Pasteur, Paris*, **78**, 177.
24. LINDEGREN, C. C. & MELLON, R. R. (1933). *Proc. Soc. exp. Biol. Med.* **30**, 110.
25. PENNINGTON, D. (1950). *J. Bact.* **59**, 617.
26. HALE, C. M. F. & BISSET, K. A. (1958). *J. gen. Microbiol.* **18**, 688.
27. BISSET, K. A. & PEASE, P. E. (1957). *J. gen. Microbiol.* **16**, 382.
28. WILSON, C. E., DONATI, E. J., PETRALI, J. P., VUICHICH, J. Y. & STERNBERGER, L. A. (1966). *Expl Molec. Path.* **Suppl. 3**, 44.
29. BISSET, K. A. (1951). *J. gen. Microbiol.* **5**, 155.
30. BERGERSEN, F. J. (1953). *J. gen. Microbiol.* **9**, 353.
31. BISSET, K. A. (1961). *J. gen. Microbiol.* **24**, 427.
32. HAYWARD, A. C., HALE, C. M. F. & BISSET, K. A. (1955). *J. gen. Microbiol.* **13**, 292.
33. DAVIS, G. H. G., BISSET, K. A. & HALE, C. M. F. (1955). *J. gen. Microbiol.* **13**, 68.
34. BISSET, K. A. & MOORE, F. W. (1949) *J. gen. Microbiol.* **3**, 387.
35. KLIENEBERGER-NOBEL, E. (1947). *J. gen. Microbiol.* **1**, 22.
36. BAIRD-PARKER, A. C. & DAVIS, G. H. G. (1958). *J. gen. Microbiol.* **19**, 446.

CHAPTER 7

Life-Cycles in Bacteria

GENERAL

The life-cycles of many bacteria are simple and direct. A cell germinates into the vegetative form and multiplies by simple, asexual fission, as well as by more complex methods (Chap. 6). The nucleus of the vegetative cell may adopt a variety of appearances. Often it is semi-permanently in the active condition, without a nuclear membrane, and with the chromatinic material in the form of chromosomes or chromosome complexes (Chap. 4).

When the food supply begins to be exhausted, and when the waste-products of the culture have accumulated to such an extent as to interfere with metabolic activity, a new generation of resting forms is produced, by a sexual or autogamous process (Chap. 5). The resting cells may be contained in elaborate fruiting bodies, or may be free. They may or may not be especially resistant. The nucleus is central and vesicular, often staining with an eccentric, chromatinic granule (Chap. 4).

In almost every case the diploid phase is short, fusion being followed by a reduction division (Chap. 5). In addition to these aspects of the bacterial life-cycle, that have already been touched upon, reproduction by motile gonidia, L-forms and other reproductive elements have been described, in various bacterial groups.

THE LIFE-CYCLE IN MYXOBACTERIA

The type of life-cycle described in the previous section is found in its most advanced form in myxobacteria. [1-6]

The unit is the swarm. When a ripe fruiting body, usually windborne, falls upon a suitable substrate, it releases the thousands of microcysts which it contains, and each of these germinates to form a vegetative bacterium, the whole constituting the swarm.

The swarm moves out over the substrate, feeding and multiplying as it goes. From time to time fruiting bodies are produced, under the influence of a specific factor elaborated by the vegetative cells. These are formed by the aggregation of vegetative bacteria, some of which are transformed into microcysts, others are sacrificed to assist in the formation of the stalk and wall of the fruiting body. The mature fruiting body, in some genera, is of great complexity, and may be borne upon a long stalk, in others it is sessile and simple in form. When ripe, the fruiting body is released from its stem and blows away in the wind. If it alights upon a suitable substrate it germinates and releases the swarm to repeat the cycle (Fig. 75).

This type of multicellularity is not peculiar to myxobacteria. It occurs also in an interesting group of organisms, the Acrasieae. In this

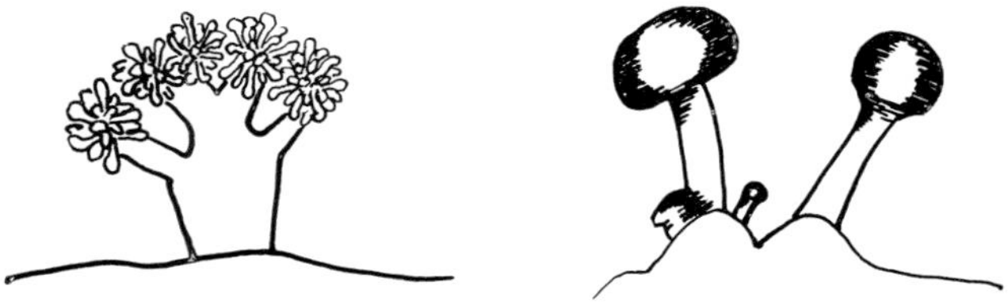

FIG. 75
MYXOBACTERIAL FRUITING BODIES
(After Krzemieniewski. Drawn from the photomicrographs)

case the unit of the swarm is an amoeboid cell instead of a bacterium, but the cycle is remarkably similar. Myxobacteria and Acrasieae are probably not in any way related, but have merely adopted a similar mode of life. They also resemble one another in being predatory upon saprophytic bacteria, in the soil, although pathogenic myxobacteria do exist, that cause diseases of fish.

THE LIFE-CYCLE IN EUBACTERIA

The condition in eubacteria and in myxobacteria is not dissimilar, although superficially it may appear to be so. Bacteriologists working with pure cultures upon otherwise sterile media can derive a false impression of the extent to which the distribution of bacteria is achieved by single cells, whether spores or otherwise, inaugurating new growth. Such conditions do not obtain in nature. The soil, which is the natural habitat of most bacteria, swarms with micro-organisms of every kind, and competition must often be too keen to permit of the immediate success of a single transplant. New substrates are more frequently introduced to micro-organisms already present, and are thus inoculated with a considerable number of spores or resting cells, derived from the last period of growth in the same area.

In effect, therefore, the unit of growth and reproduction, in eubacteria as in myxobacteria, is the swarm or bacterial culture. The culture grows and ages, physically and physiologically, exactly like a multicellular organism, and differs from one mainly in the lack of specialisation of its component cells. In myxobacteria some degree of specialisation has been achieved in the formation of the various parts of the fruiting body, but in eubacteria all cells have an equal reproductive potential.

The fact that sporing bacilli have a life-cycle has been known for so long that it tends to be ignored in any discussion of bacterial life-cycles. The endospore is uniquely resistant, but since the power to resist heat and chemicals cannot be of any service to a soil-dwelling micro-organism, it is probable that these characters are by-products of a reduction in size and weight, that makes the spore a more efficient distributive agent, [7] by the elimination of some of its water-content. [8] That the endospore is actually adapted to aerial dispersion, rather than to resistance, is verified by the recent observations of Rode [9, 10] and his collaborators, who showed that some spores possess relatively long, feathery appendages, that can be likened to the parachutes of certain seeds (Figs 26 & 27). It is possible that the deep sculpturing of spore coats [11] may have a similar effect to a certain degree (Fig. 14).

The spore, however, is unique only in its powers of resistance. Many bacteria have resting cells or microcysts, rather like those of myxobacteria, [12] although not enclosed in a fruiting body. They may be quite distinctive in appearance [13] so that they provide valuable (but quite unused) diagnostic characters (Fig. 40). Some have been known for many years (Chaps 4 & 5), and one of these, the cyst of Azotobacter, is very distinctly spore-like in its development and structure, as seen in electron micrographs of ultra-sections. [14]

LIFE-CYCLES OF HIGHER BACTERIA

In streptomyces a truly multicellular organism is formed, and thus the problem of distribution entails the liberation of free, reproductive units, small airborne spores. Streptomyces resemble moulds in their general form, and this resemblance extends to their mode of reproduction.

The spore alights upon a suitable substrate, and germinates to form the primary mycelium. Specialised branches arise upon the primary mycelium and initiate the aerial hyphae which bear chains of spores. These are often provided with hairs or spines [15] (Fig. 92) that must be presumed to improve their efficiency in aerial distribution.

During its lifetime the organism or colony produces spores continuously, while conditions are suitable. In eubacteria, upon the other hand the reproductive cells arise only in an ageing culture, when almost all may be thus transformed.

The true Actinomyces which are parasitic and microaerophilic, also have a complete life-cycle with a primary and secondary mycelium, but the spores are borne singly and in relatively small numbers. This may represent a degenerate condition. [16] Similarly degenerate traces of a life-cycle are found in other parasitic, filamentous bacteria, for example Leptotrichia. [16, 17] The free-living Micromonospora also shows signs of degeneracy by comparison with Streptomyces and somewhat resembles Actinomyces. [18] It is a more specialised organism than Streptomyces, and tends to live under less aerobic conditions, for example, in compost. This may account for the difference. The life-cycles that have been claimed to exist in mycobacteria are of a rather different character. The growth-cycles have been described in the last chapter, and the problem of filterable or sub-cellular forms will be discussed in the section on Gonidia, below.

MOTILE DISTRIBUTIVE BACTERIA

Many aquatic bacteria have stages in their life-cycles involving motile, full-sized distributive cells, that may be markedly different from the sessile vegetative forms. In chlamydobacteria, such as the filamentous, iron bacterium *Sphaerotilus discophorus*, the vegetative cells grow as a long filament, surrounded by a gelatinous sheath containing colloidal iron. At the ends of the sheath the cells are transformed into flagellated swarmers, which swim off to found new, sessile, filamentous colonies. It is unfortunate that the cytology of this interesting process has never been properly described. The vegetative cells are quite similar to those of eubacteria, but the minute structure of the swarmers has never been recorded. Release of cells from the sheath was described in 1870 by Cohn [19] and by many subsequent workers. Pringsheim [20] showed that some of these were motile, and others non-flagellate. Motile cells have been observed by several recent investigators [21, 22] but it is not certain whether these are simply released by disruption of the sheath, or

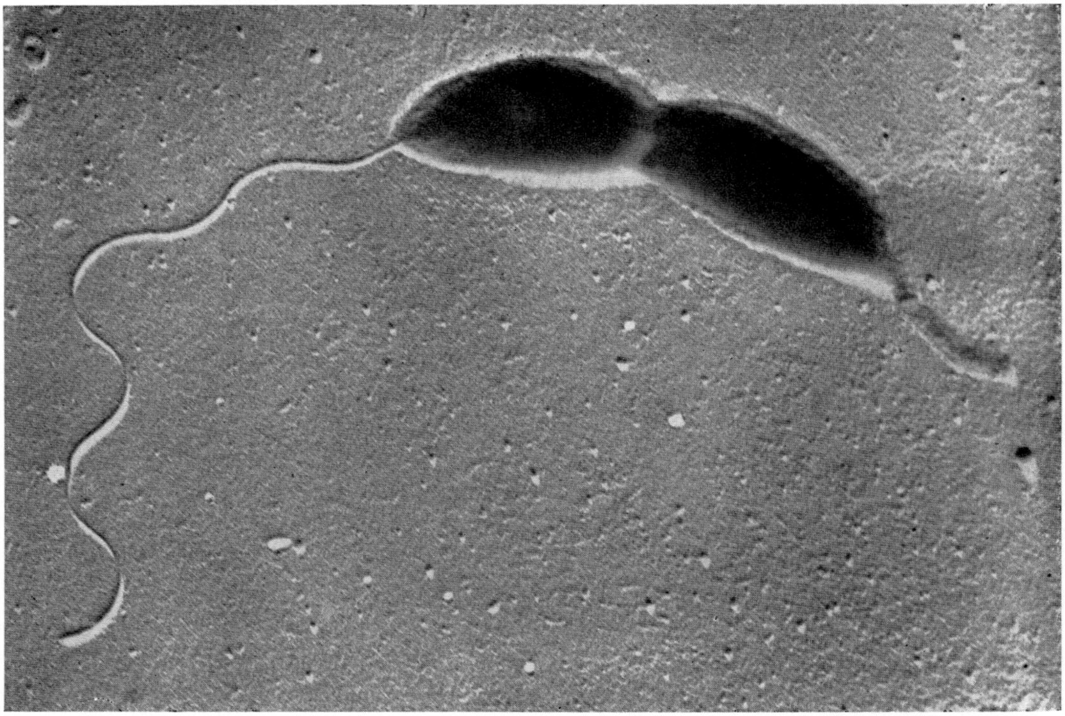

(*Electron micrograph by Dr Phyllis E. Pease*)

FIG. 76

THE LIFE-CYCLE OF *CAULOBACTER*

The life-cycle of the true, stalked caulobacteria provides an example of the alternation of sessile and motile generations, such as is commonly found in other biological groups. In Fig. 76 the entire cycle is seen foreshortened. The stalked cell is producing a flagellate daughter cell. Fig. 77 shows the complete life-cycle. The stalked generation (1) is shown in process of division. In (2) the flagellate daughter cell is shown. In (3) are two flagellate cells in the process of becoming sessile. In the upper example the stalk has already developed, but the flagellum (slightly outlined for clarity) is retained. In the lower example the stalk is in an early stage of development. Electron micrographs, gold-shadowed. Fig. 76 and Fig. 77 (1) ×20,000; Fig. 77 (2), (3) ×15,000.

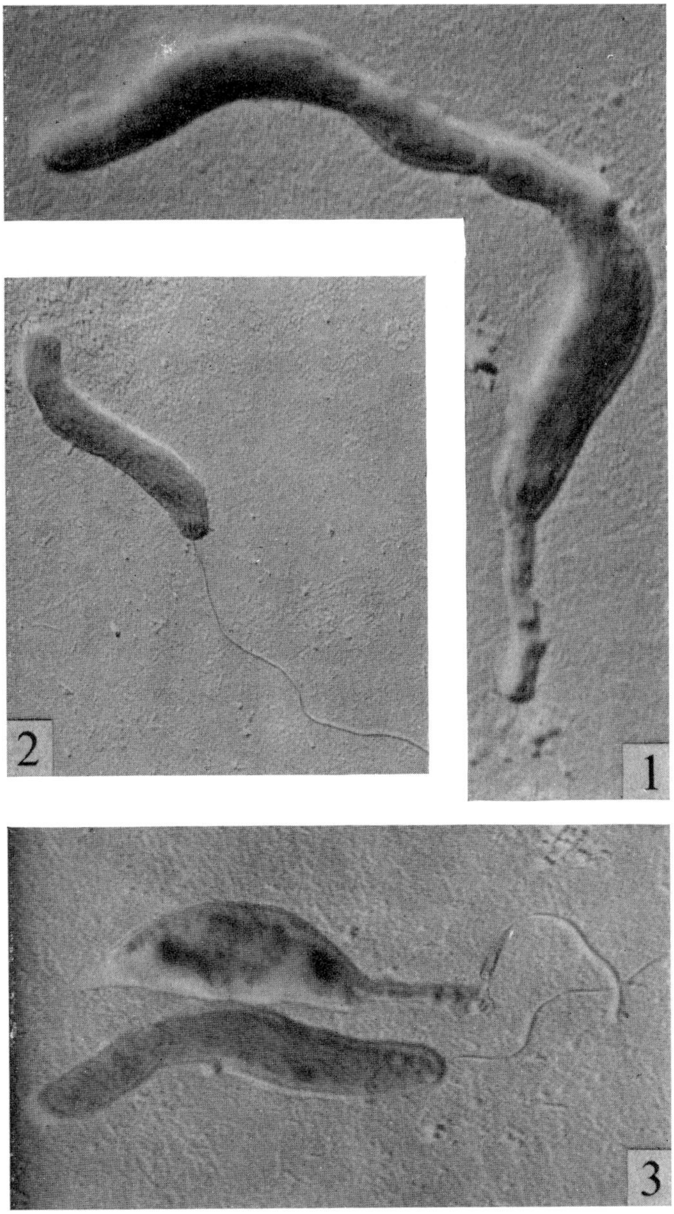

FIG. 77

whether they are special swarm cells or gonidia, deriving from the disruption of a mother-cell.

Aquatic actinomycetes with motile spores have been described (23) but it is not at all easy to learn, from the evidence, whether some of these may not be forms of chlamydobacteria or unusually elaborate caulobacteria.

Caulobacter has a characteristic, simple cycle (Fig. 76 & 77). The sessile, stalked bacterium produces a succession of motile, polar-flagellate daughter cells that swim actively until they find a suitable attachment upon which to settle, form a stalk, and produce a new, sessile generation. (24, 25) The effect is exactly as though a sessile, asexual generation were producing a series of motile buds. If this type of organism possesses a sexual stage, it is in the motile cells that it must be sought.

The hyphomicrobial caulobacteria (Fig. 86) produce a series of buds on a branched-stalk and also have occasional motile, flagellated cells, that may serve a distributive function. (26) These organisms are probably not closely related to Caulobacter, and may be a phase of growth of other, more typical bacteria. (26)

A little-known bacterium that also has the power of budding-off motile cells is Dermatophilus. This veterinary parasite has numerous septa and cross-walls, dividing the filaments in all directions. Motile buds can thus be produced from any part of it. (27, 28)

GONIDIA AS A STAGE IN THE BACTERIAL LIFE-CYCLE

Numerous descriptions have been published of reproduction in bacteria, as in other flagellate protista, by the liberation of tiny,

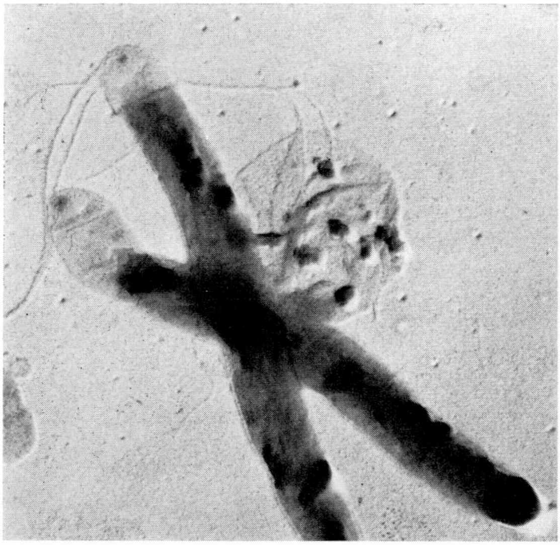

(Reproduced from the Giornale di Microbiologia)

FIG. 78
CONJUGATION IN *SPIRILLUM*

Electron micrograph of two conjugating (?) spirilla, bearing a cyst with small granules. ×15,000.

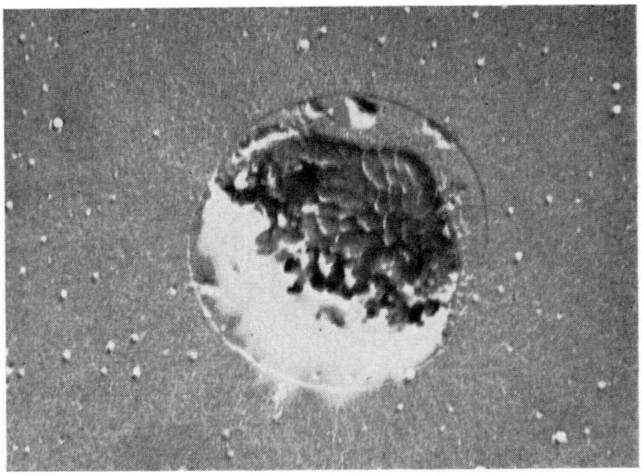

(Electron micrograph by Dr Phyllis Pease)

FIG. 79
CONJUGATION IN *SPIRILLUM*

Reproductive cyst of a spirillum (as in Figure 78). The granules are developing into small bacteria. ×15,000.

filterable granules or gonidia, but few have been accompanied by sufficiently detailed, cytological information. Nor have research workers or writers of reviews [29, 30] really attempted to distinguish these from the better-known L-forms, that are discussed in the next paragraph.

True gonidia, that is to say, complete but small, actively motile, reproductive cells, are most readily demonstrated in the spirilla and spirochaetes [29, 30, 31] (Figs 78 & 79), in sporing bacilli [32, 33, 34] (Figs 80 & 81), and in the nitrogen-fixing bacteria Azotobacter and Rhizobium [35-40] (Figs 82 & 83). The general phenomenon, and indeed, most of the cytological details in all these cases, have long been known, but have been described in a rather unconvincing manner, and, in the case of Rhizobium, not entirely accurately. It was supposed that large 'barred' bacilli fragmented to form small coccoid swarmers, each dark bar representing the genesis of a single swarmer. In fact, the process is alike in all genera. The tiny, polar-flagellated gonidia develop within the lumen of large mother cells, and are released by rupture of the cell wall. The mother cells of Rhizobium are divided by basophilic septa, probably secretory in function, and these septa are the 'bars' of the earlier account.

Many early accounts of gonidial reproduction refer to spore-bearing bacilli, and of these the paper published by Allen *et al.* [32] is the most detailed and convincing. These workers described the occurrence of small, refractile granules in the cytoplasm of the bacillus. The granules appeared to reproduce by fission, and were liberated from the cell and transformed into small rods which grew up into normal bacilli. This was, in general, confirmed by Bisset, [33, 34] who also showed that the gonidia possessed a much thinner cell wall than the vegetative cells, and suggested that this might be regarded as evidence of a relationship between these structures and the cell-wall-free L-forms (see below) (Fig. 84).

The existence of a granular reproductive phase in spirochaetes has long been a subject of controversy. Recent studies have confirmed, however, that members of this group may reproduce both by transverse fission and by the formation of large cysts, usually at the end, but occasionally in the middle of the organism. These cysts contain several small spirochaetes, sometimes in a granular form. [29, 41] In the experience of the author the cysts stain more deeply than the vegetative spirochaetes with basic dyes, and thus probably have an increased nucleic acid content. There is evidence that the granules or gonidia are more resistant than the vegetative spirochaetes, and the latter may survive adverse conditions in this form. The details of the processes of formation and germination have not been described nor is it known whether a sexual process is involved. Ultra-sections of the developing sacs show that the gonidia of spirochaetes, like those of sporing bacilli, have a reduced cell wall. [42]

Much emphasis has always been placed upon the ability of gonidial forms to pass antibacterial filters. The swarmers of sporing bacilli and nitrogen-fixing genera, at least, can perform this test successfully, although they are complete, if small bacteria [31, 43] and the buds of hyphomicrobia are equally small (Fig. 86).

G-FORMS, L-FORMS AND MYCOPLASMA

Bacterial L-forms are essentially vegetative bacteria that have lost part of their cell en-

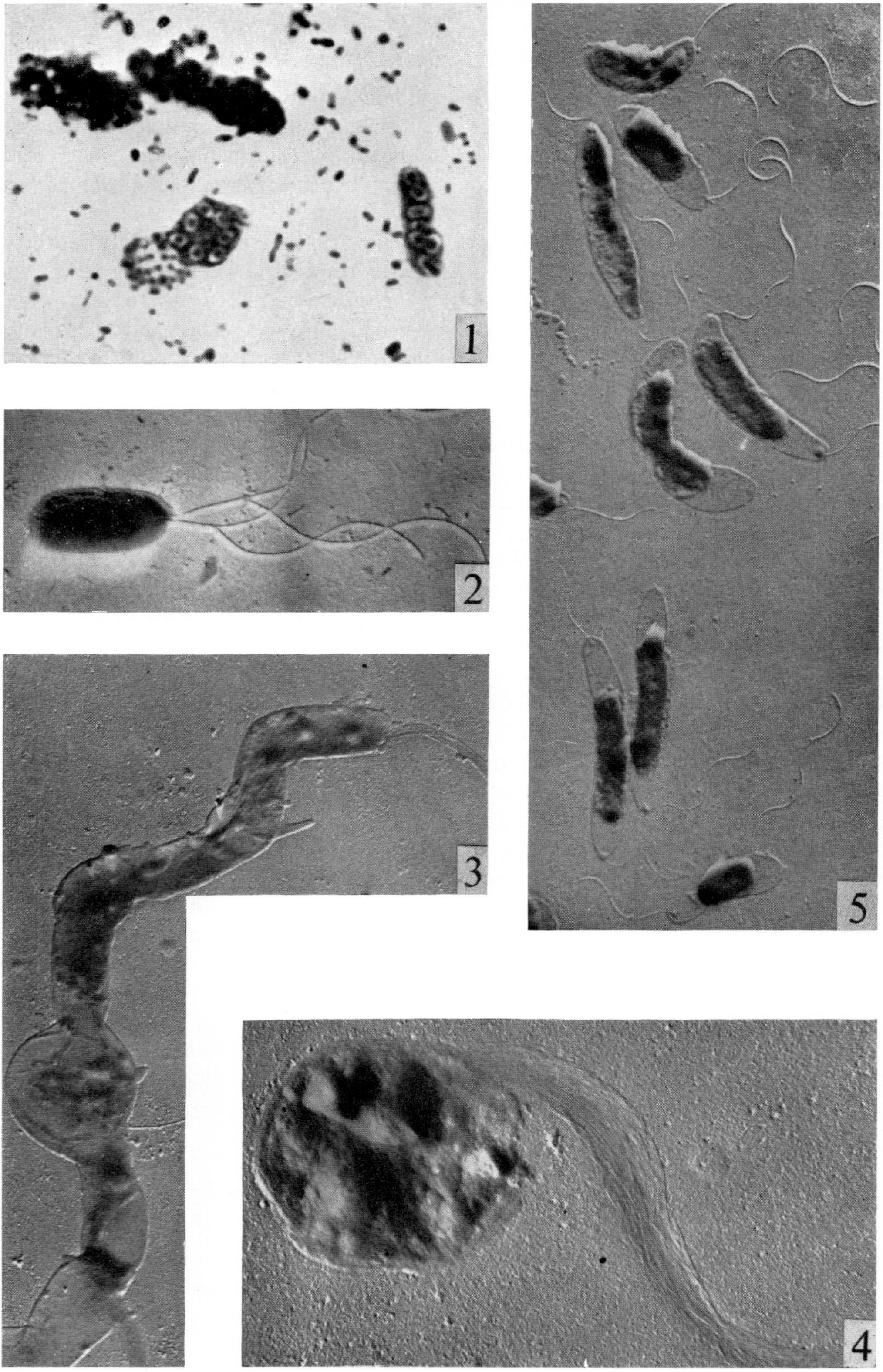
1
2
3
4
5

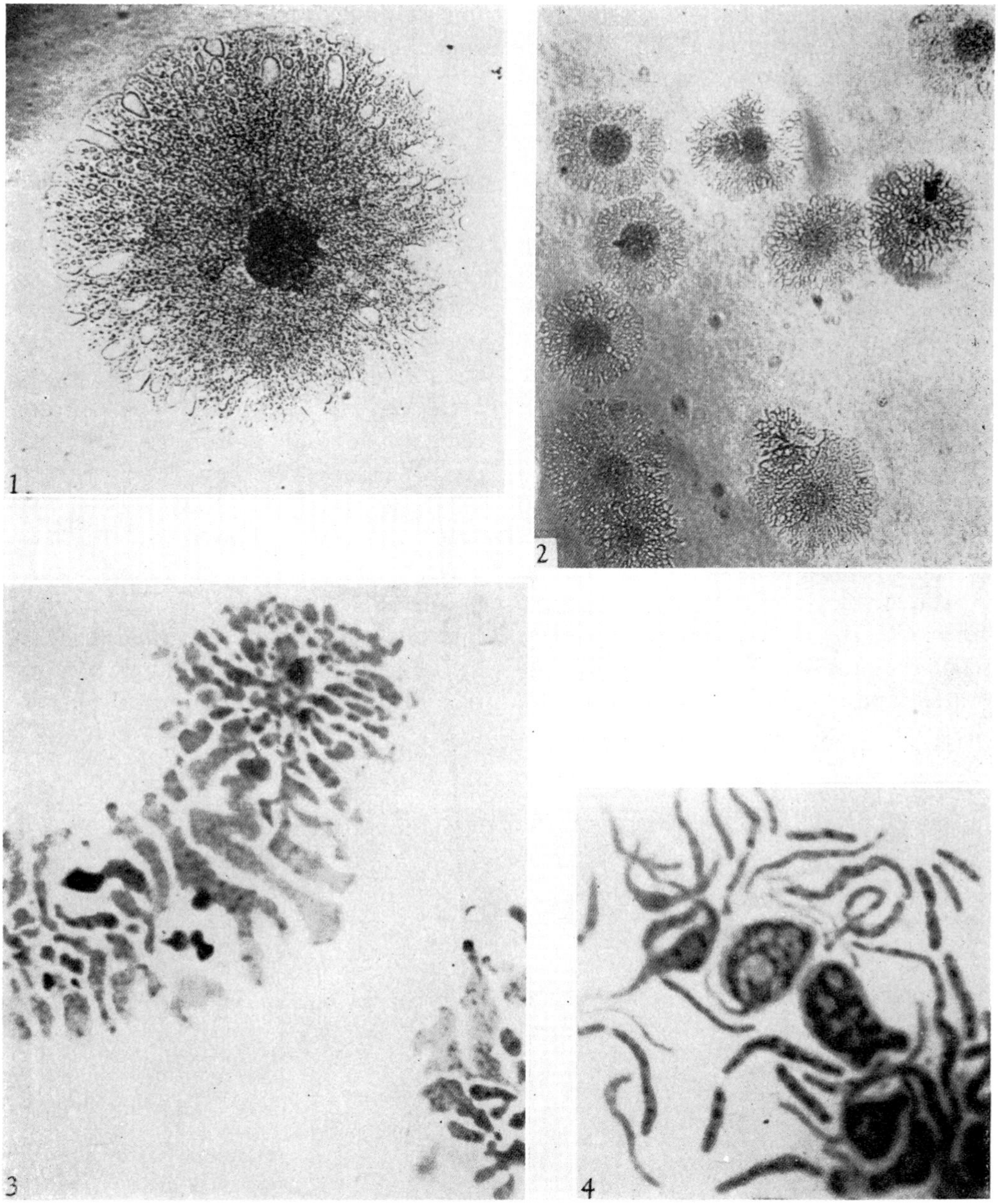

(*Photomicrographs by E. Klieneberger-Nobel. Reproduced from the Journal of General Microbiology*)

FIG. 85

THE L-STAGE IN THE BACTERIAL LIFE-CYCLE

(1), (2) Colonies of the L-form of *Fusiformis necrophorus* ×200. This gonidial stage reproduces for some time without reverting to the bacterial form.

(3) A colony of the L-form, fixed and stained, *in situ* ×2000.

(4) *Fusiformis necrophorus*, the bacterial form, showing reproductive swellings. Giemsa ×3500.

tain its shape. [46] It has been suggested that penicillin encourages the synthesis of a defective wall, which is apt to be destroyed subsequently by lytic enzymes. [51] The completely cell-wall-free stages may break down into the very small bodies already referred to. [44]

In the larger forms, at least, the nucleus is unaltered from its state in the original bacillus [52] but it is usually attached to the membrane and rests at the circumference of the cell. This was observed some years ago by Jeynes [53] and it is now known to be due to the attachment of the nucleus to the mesosome [52] which, being a diverticulum of the cell membrane, is liable to be stretched out flat when the cell loses its retaining wall (Chaps 3 and 4). This eccentric nucleus can be seen in electron micrographs of the larger, but not the smaller bodies composing some L-cultures. [44] The condition of the nuclear material in the small bodies has never been demonstrated, and classical cytological studies, for example, those of Tulasne, [54] suggest that the nuclei of the large bodies break up into very small fragments, when they are thus transformed.

When L-forms appear naturally, as in the classical cases of *Streptobacillus moniliformis* [47] or Proteus, [54] they arise as swellings on filamentous bacilli. These swellings may be transformed into L-colonies (Fig. 85) consisting of several large or many small bodies (often with a halo of the latter surrounding the former), or they may simply break up into large and small protoplasts, that divide and reproduce by a budding type of fission. [53] They return to the bacterial phase, simply by growing a cell wall.

It has been claimed that the swellings whereby the L-cycle is initiated occur at the points of fusion of two Proteus swarms, and that a sexual fusion is entailed [55, 56] (Chap. 5). This may be true, although it is certainly true also that various inimical agencies can

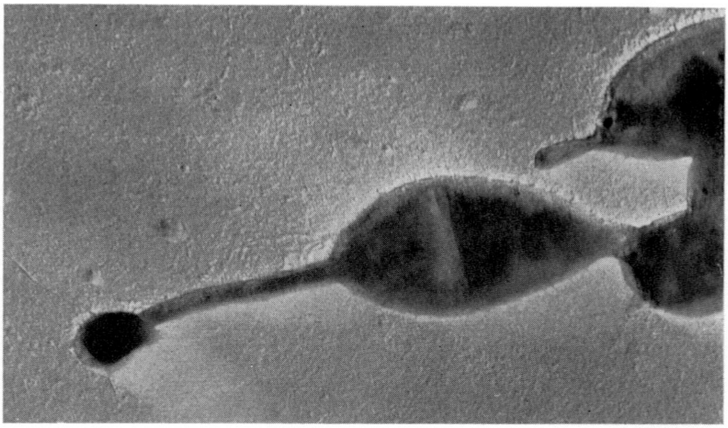

(*Electron micrograph by Dr M. H. Jeynes*)

FIG. 86
REPRODUCTION BY BUDDING

Hyphomicrobium sp. This organism reproduces by budding, and each bud is comparable in size to the gonidial granules of other bacteria. $\times 15,000$.

bring about L-form transformation without the initial intervention of contact with another culture. This does not mean that fusions may not occur at the same time, but the evidence is scanty (Fig. 67).

SUMMARY

The type of life-cycle that is seen in its most perfect form in myxobacteria is also common to most other bacterial groups.

In the resting stage, a group of spores or microcysts is transplanted upon a fresh medium, and germinates to produce the vegetative culture or swarm, which is the reproductive condition. When the substrate is exhausted the vegetative cells undergo a sexual process to produce the resting stage, and remain in that condition until again transplanted, or until the food supply is renewed.

The resting stage may be a resistant spore, or may not be markedly resistant, except to inanition. The resistance of endospores is due to their reduced water content, and is probably accidental; their main purpose is distribution. In the case of myxobacteria the microcysts are contained in elaborate fruiting bodies.

Sessile bacteria, the mycelium-forming streptomyces, filamentous chlamydobacteria and stalked caulobacteria can only be distributed by the agency of free, reproductive units. Streptomyces produce aerial spores in large numbers, and the aquatic chlamydobacteria and caulobacteria produce motile swarm cells.

Many bacteria can produce very small gonidia from which typical bacteria are regenerated. Such gonidia are found in their most perfect form in the life-cycles of Rhizobium, Azotobacter and certain spiral bacteria.

L-forms are vegetative bacteria with a deficient cell wall, but they also produce smaller gonidium-like stages, some of which have lost almost all resemblance to bacteria.

REFERENCES

1. Anscombe, F. J. & Singh, B. N. (1948). *Nature, Lond.* **161**, 140.
2. Beebe, J. M. (1941). *J. Bact.* **42**, 193.
3. Garnjobst, L. (1945). *J. Bact.* **49**, 113.
4. Klieneberger-Nobel, E. (1947). *J. gen. Microbiol.* **1**, 33.
5. Krzemieniewski, H. & S. (1926). *Acta Soc. Bot. Pol.* **4**, 1.
6. Lev, M. (1954). *Nature, Lond.* **173**, 501.
7. Bisset, K. A. (1950). *Nature, Lond.* **166**, 431.
8. Ross, K. F. A. & Billing, E. (1957). *J. gen. Microbiol.* **16**, 418.
9. Rode, L. J., Crawford, M. A. & Williams, M. G. (1967). *J. Bact.* **93**, 1160.
10. Pope, L., Yolton, D. & Rode, L. J. (1967). *J. Bact.* **94**, 1206.
11. Bradley, D. E. & Williams, D. J. (1957). *J. gen. Microbiol.* **17**, 75.
12. Bisset, K. A. (1949). *J. Hyg., Camb.* **47**, 182.
13. Bisset, K. A. (1950). *J. gen. Microbiol.* **4**, 413.
14. Bisset, K. A. (1967). *J. gen. Microbiol.* **48**, 25.
15. Baldacci, E., Gildardi, E. & Amici, A. M. (1956). *G. Microbiol.* **6**, 512.
16. Bisset, K. A. & Davis, G. H. G. (1960). *The Microbial Flora of the Mouth.* London: Heywood.
17. Baird-Parker, A. C. & Davis, G. H. C. (1958). *J. gen. Microbiol.* **19**, 446.
18. Bisset, K. A. (1957). *J. gen. Microbiol.* **17**, 562.
19. Cohn, F. (1870). *Beitr. Biol. Pfl.* **1**, 108.
20. Pringsheim, E. G. (1949). *Biol. Rev.* **24**, 200.
21. Mulder, E. G. & van Veen, W. L. (1963). *Antonie van Leeuwenhoek*, **29**, 121.

22. Doetsch, A. (1966). *Arch. Mikrobiol.* **54**, 46.
23. Higgins, M. L., Lechevalier, M. P. & Lechevalier, H. A. (1967). *J. Bact.* **93**, 1446.
24. Bowers L. E., Weaver, R. H., Grula, E. A. & Edwards, O. F. (1954). *J. Bact.* **68**, 194.
25. Houwink, A. L. (1955). *Antonie van Leeuwenhoek*, **21**, 49.
26. Bisset, K. A. (1961). *J. gen. Microbiol.* **24**, 427.
27. Bisset, K. A. & Thompson, R. E. M. (1956). *J. Path. Bact.* **72**, 322.
28. Gordon, M. A. (1954). *J. Bact.* **88**, 509.
29. Hampp, E. G., Scott, D. B. & Wyckoff R. W. G. (1948). *J. Bact.* **56**, 755.
30. Pease, P. E. (1956). *J. gen. Microbiol.* **14**, 672.
31. Bisset, K. A. & Hale-McCaughey, C. M. F. (1967). *G. Microbiol.* **15**, 119.
32. Allen, L. A., Appleby, J. C. & Wolf, J. (1939). *Zentbl. Bakt. ParasitKde, II*, **100**, 3.
33. Bisset, K. A. & Hale, C. M. F. (1963). *J. gen. Microbiol.* **31**, 281.
34. Bisset, K. A. (1966). *G. Microbiol.* **14**, 5.
35. Jones, D. H. (1920). *J. Bact.* **5**, 325.
36. Löhnis, F. (1921). *Mem. natn. Acad. Sci.* **16**, 1.
37. Thornton, H. G. & Gangulee, N. (1926). *Proc. R. Soc.* **B**, **99**, 427.
38. Bisset, K. A. & Hale, C. M. F. (1951). *J. gen. Microbiol.* **5**, 592.
39. Bisset, K. A. (1952). *J. gen. Microbiol.* **7**, 233.
40. Bisset, K. A. & Hale, C. M. F. (1953). *J. gen. Microbiol.* **8**, 442.
41. DeLamater, E. D., Wiggall, R. H. & Haanes, M. (1950). *J. exp. Med.* **92**, 239 & 247.
42. Bladen, H. A. & Hampp, E. G. (1964). *J. Bact.* **87**, 1180.
43. Lawrence, J. C. (1955). *Nature, Lond.* **176**, 1033.
44. Pease, P. E. (1957). *J. gen. Microbiol.* **17**, 64.
45. Pease, P. E. (1957). *G. Microbiol.* **3**, 44.
46. Dienes, L. & Bullivant, S. (1968). *J. Bact.* **95**, 672.
47. Klieneberger, E. (1935). *J. Path. Bact.* **40**, 93.
48. Hadley, P., Delves, E. & Klimek, J. (1931). *J. infect. Dis.* **48**, 1.
49. Burgess, E. (1924). *Ceylon J. Sci.* p. 33.
50. Brenner, S. *et al.* (1958). *Nature, Lond.* **181**, 1713.
51. Park, J. T. (1968). In *Microbial Protoplasts, Spheroplasts and L-forms. A Symposium*, ed. Guze, L. B. p. 52. Baltimore: Williams and Wilkins.
52. Ryter, A. & Landman, O. E. (1968). In *Microbial Protoplasts, Spheroplasts and L-forms. A symposium*, ed. Guze, L. B. p. 110. Baltimore: Williams and Wilkins.
53. Jeynes. M. H. (1961). *Expl Cell Res.* **24**, 255.
54. Tulasne, R. (1949). *C. r. Séanc. Soc. Biol.* **143**, 286.
55. Dienes, L. (1946). *Cold Spring Harb. Symp. quant. Biol.* **11**, 51.
56. Dienes, L. & Bullivant, S. (1967). *Ann. N.Y. Acad. Sci.* **143**, 719.

CHAPTER 8

Macroformations

THE MYXOBACTERIAL FRUITING BODY AND SWARM

The most perfect and elaborate multicellular structures formed by bacteria are the fruiting bodies of myxobacteria. Other macroscopic formations, however elaborate, are little but the result of the reaction between the growth potential of the organism and the physical restraint of the environment, and slight variations in the latter may affect the result to an apparently disproportionate extent. The fruiting body, however, although it may be prevented by unsuitable conditions from forming at all, is otherwise independent of small environmental changes, and is characteristic of the species [1, 2] (Fig. 75). The co-ordination of cellular activity which initiates the formation of the fruiting body is stimulated by a specific substance, analogous to a hormone that diffuses from the vegetative cells. [3]

The great majority of the cells are transformed into typical microcysts and thus survive. Some are embodied in the stalk and envelope, and are sacrificed. Little is known of the mode of formation of these structures. It has been stated that the cells which take part in their formation are cemented together by dried mucus, but this appears to be mere supposition. The physical properties of the envelope vary considerably in different species, and it may even be entirely absent. The envelope varies especially in its physical strength and in its resistance to water. The fruiting bodies of some species of myxobacteria burst open as soon as they are wetted. Others remain intact. This characteristic has been considered to be of taxonomic value by some botanists, but there is no reason to believe that it indicates biological relationship. Whether the variation is due to differences in structure and composition of the envelope is not known.

In the vegetative stage also, myxobacteria give the impression of a degree of organisation far beyond that of other bacteria. The swarm moves outwards from the centre of germination in a regular fashion, following the lines of physical stress in the substrate. [4] It concentrates in chosen areas, converging towards the incipient fruiting body, and piling up, the bacteria crawling over each other, to encyst in an elevated mass. The appearance of ordered purpose is most remarkable in so lowly an organism.

THE SWARM OF PROTEUS

Some degree of cell-specialisation and organisation in the swarm of Proteus is indicated by several workers. [5, 6] It is suggested that the swarm commences its activity when an initial generation of large cells has produced a sufficient concentration of metabolites to provide the energy for swarming. [7] The swarm cells are filamentous and have a very large number of flagella. Some of them are also thicker than the vegetative cells, having twice or even four times the diameter, according to Preusser. [8] They move out rapidly over the substrate until their reserve of energy is exhausted, and then rest until it becomes possible to repeat the process. Although this phenomenon is most advantageously seen upon the surface of an agar plate, there is no reason to believe that it does not occur in nature. Proteus may accordingly be regarded as having achieved a minor degree of cellular specialisation. It is also a temporary specialisation, because the swarm filaments, which are distributive in function, are the descendants as well as the parents of the 'somatic' cells.

CHLAMYDOBACTERIAL AGGREGATES

As far as is known, the condition in chlamydobacteria, typified by the filamentous, iron bacteria, resembles that in Proteus. [9, 10, 11] The aggregates of gelatinous sheaths, the by-product of metabolism, are devoid of structural specialisation, except in so far as some are dead casts left behind by the cells which were responsible for their formation, whereas others are inhabited, and

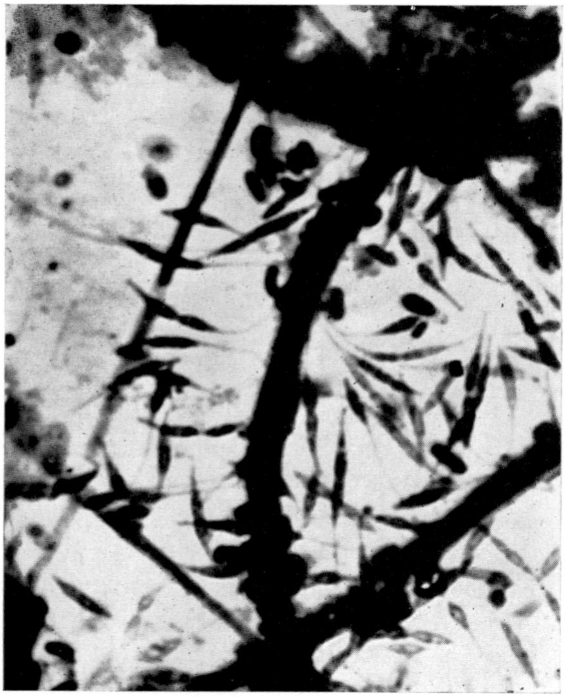

FIG. 87
CAULOBACTERIAL AGGREGATES

A colony of *Caulobacter* attached to filaments of *phaerotilus*. Each cell has an independent very fine stalk. (Unstained) ×1500.

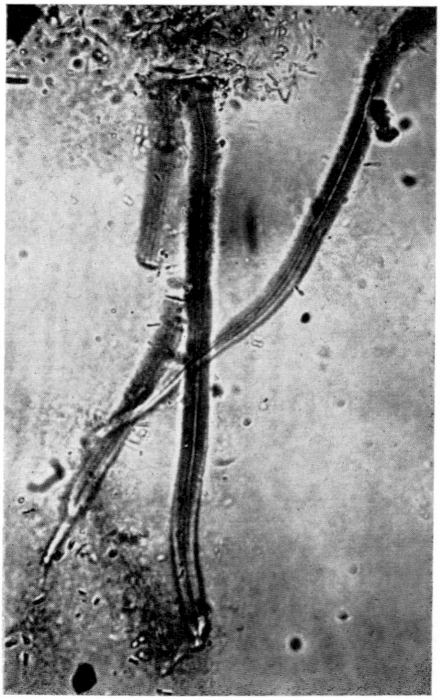

FIG. 88
CHLAMYDOBACTERIAL AGGREGATES

A group of filaments of the chlamydobacterium *Sphaerotilus discophorus*. The sheath is composed of colloidal ferric hydroxide. The older portions are the thickest and may be vacant. The growing filaments protrude from the sheath at the thinner end. (Unstained) ×1000.

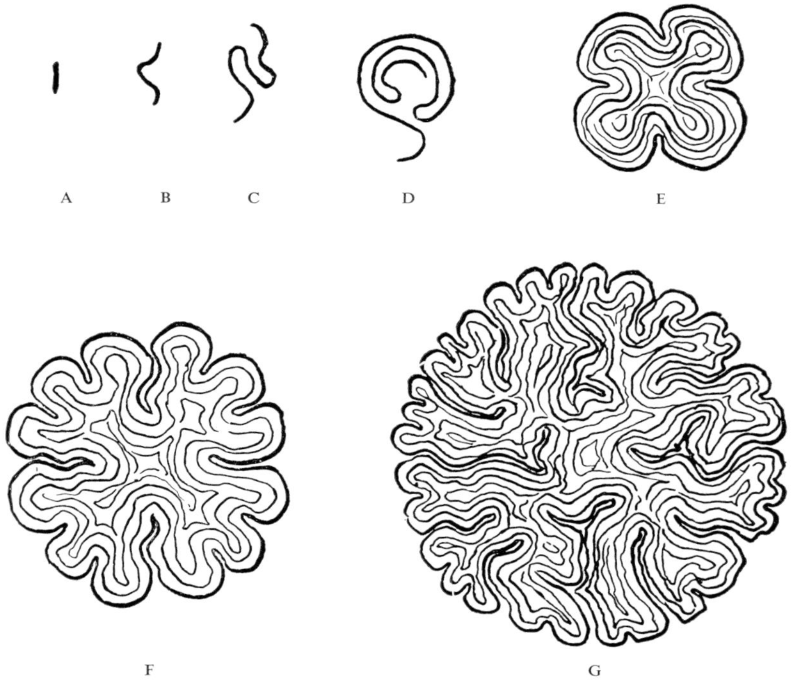

(Reproduced from the Journal of Pathology and Bacteriology)

FIG. 89
STAGES IN THE GROWTH OF A MEDUSA-HEAD COLONY

A. Original bacillus.
B, C. Elongation and looping.
D. Primary coil.
E. Further growth, against the friction of the medium causes infolding of the coil.
F. Secondary infolding of the coil.
G. Continued growth of all parts of the colony causes complex folding and convolution.

still increasing in size. The holdfasts by which the ends of some types of filament are attached to the substrate are a possible exception to this rule. Those bacteria in which holdfasts are most strongly developed are often classed as caulobacteria for precisely this reason (Figs 87 & 88).

After a period of vegetative growth, motile swarm cells are produced and swim away to form new colonies of filaments. There is no evidence that all the cells comprising the sessile filaments may not be equally capable of transformation into swarmers; and indeed, what little is known of their cytology suggests a close resemblance to eubacteria, in this and other respects (Chap. 7).

The aggregates formed by stalked caulobacteria may consist of small clumps attached to a single point upon the substrate, or of free colonies, in the form of rosettes or ribbons, composed of numerous bacteria joined together by their stalks [10] (Fig. 87).

THE MEDUSA-HEAD COLONY

In most bacteria the colony, however complex its structure, is an accidental growth, each cell of which is equivalent to all the others. The colony does not alternate with the swarm, but is itself the swarm, or vegetative mass. Colonies are not confined to conditions of artificial culture, although there they appear in their most obvious form.

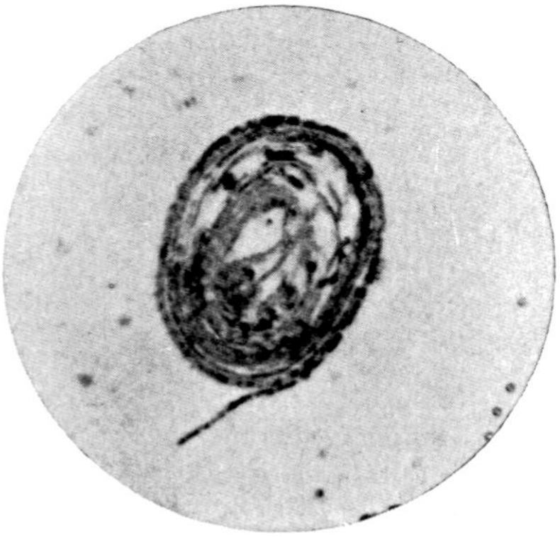

(Reproduced from the Journal of Pathology and Bacteriology)

FIG. 90
GROWTH OF A ROUGH COLONY

Primary coil of a rough colony. *Shigella flexneri*, impression preparation ×700.

The type of bacterial colony whose structure was first recognised, although not understood, was the so-called 'medusa-head' colony of the anthrax bacillus. This formation which consists of long, coiled bacillary threads, is common to all bacteria of what we have termed rough morphology. In the case of the anthrax bacillus, and similar, large, spore-bearing bacilli, it is easily seen with a hand lens, whereas the much smaller size of, for instance, lactobacilli, renders it less obvious, so that its presence, except in these large genera, was unsuspected. [12-15]

Although it has long been known that the virulent anthrax bacillus possesses this type of colony, whereas the smooth colonies of the avirulent anthrax vaccine were composed of individual bacilli, the converse corollary that the respectively virulent and avirulent smooth and rough colonies of Bacteriaceae might possess the same type of structure in each case, escaped attention until much later.

The structural complexity, even beauty, of the medusa-head colony has tended to produce the impression that there exists an intrinsic tendency towards the formation of the structure, as in the case of a true, multicellular organism. This is not so. A rough bacillus, growing upon a frictionless surface, would produce a straight, or slightly spiral thread of indefinite length. Upon an agar plate, however, after the thread has grown a short way across the surface, its rigidity is not sufficiently great to permit it to extend further in a straight line. It therefore kinks, and because of its slightly spiral growth, tends to form flat coils upon the surface of the plate. All portions of this primary coil are growing simultaneously, and to accommodate the growth it produces secondary loops and coils, and upon them still further and more complex convolutions. The outer portions of the colony lie flat upon the medium, and the internal coils overlie one another to a small extent (Figs 15, 89 & 90).

The appearance and complexity of the colony vary with the rigidity of the bacterial thread, and the resistance of the surface of the medium. Bacteria of rough morphology vary considerably in rigidity between such extremes as *Bacillus mycoides*, which is so rigid that it seldom produces any structure

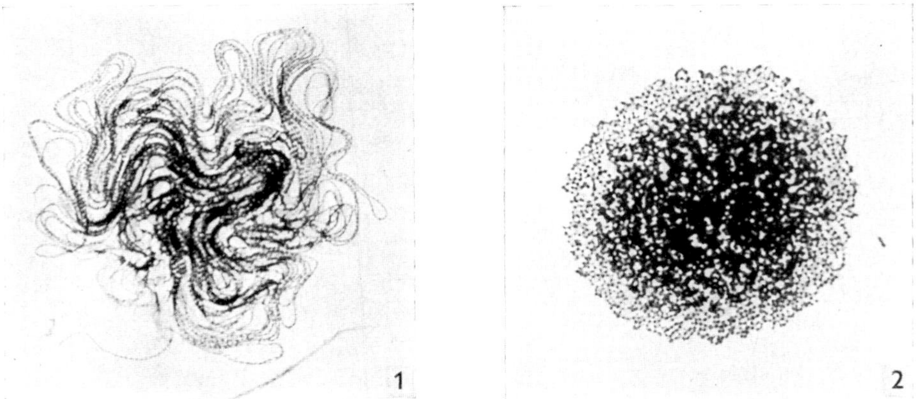

(*Reproduced from the Journal of General Microbiology*)

FIG. 91
COLONIES OF STREPTOCOCCI

1. Long-chained 'rough' colony.
2. Short-chained 'smooth' colony.
   Impression preparations ×300.

more complex than a primary coil, and grows out as long threads across the agar, and on the other hand, those rough Bacteriaceae whose colonies are almost indistinguishable from smooth variants. The long-chained type of Streptococcus also produces a variety of medusa-head colony (Fig. 91).

The structure of these colonies may be entirely disguised by the production of mucoid capsular material. The anthrax bacillus produces its polypeptide capsule under suitable cultural conditions, of which CO_2 tension is one of the most important. The phenomenon of Smooth→Rough variation in pneumococci is entirely concerned with the production or loss of polysaccharide capsular material, concealing or revealing the rough appearance of the colony.

COLONIES OF STREPTOMYCES

The colonies of streptomyces may be considered as single, multicellular organisms, in which the functions of the various types of component cell are almost completely specialised. The colony consists of a vegetative, haploid, primary mycelium, upon which arises a reproductive, diploid, secondary mycelium, of somewhat larger diameter.[16-19] The condition is similar to that which obtains in higher fungi. The spores are specialised distributive cells, often provided with hairs or spines to increase their sensitivity to air currents [17] (Fig. 92). A single spore may give rise to the complete colony. Although fragments of mycelium will grow if they are transplanted, the colony is essentially a unit that remains fixed upon its substrate, so long as the food supply permits, and it reproduces solely by the release of spores. The cells of the mycelium perish when the food supply is exhausted or when conditions become unsuitable for growth.

Thus, in bacteria, the type of multicellularity represented by the swarm finds its greatest perfection in myxobacteria, the

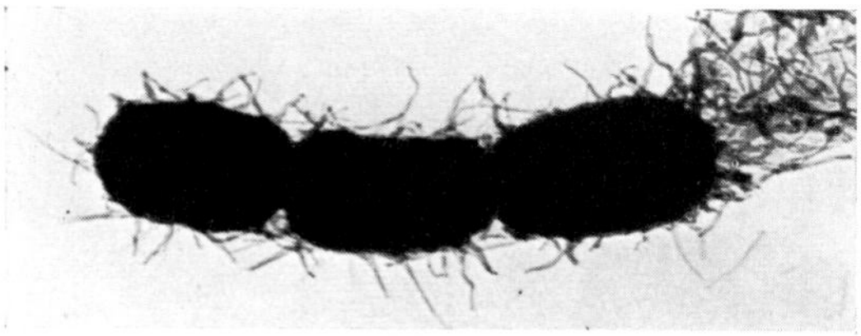

(*Electron micrograph by L. O. White*)

FIG. 92

Spores of *Streptomyces* sp. showing spines.

sessile colony in the streptomyces. In each the mode of reproduction and distribution is admirably designed to the case.

SUMMARY

Bacteria may form multicellular structures of several different types. In some of these there is a certain degree of specialisation of function between the constituent bacteria, in others little or none. In the myxobacterial fruiting body the majority of bacteria are transformed into reproductive microcysts, the minority become 'somatic' cells in the stem or wall, and are sacrificed. This is also true, in a greater degree, of streptomyces. The main body of the colony is sessile and vegetative. the spores are borne upon a special reproductive mycelium. A similar specialisation of function divides the sessile and motile cells of chlamydobacteria and caulobacteria, and the swarming and non-swarming units of Proteus.

Streptomyces colonies are single units, arising from a single spore. In eubacteria and myxobacteria the unit is the swarm or vegetative culture. The myxobacterial fruiting body is a device for ensuring the distribution and survival of swarms, as such. In the eubacteria distribution may be of single cells or of portions of swarms, usually the latter. Eubacterial colonies have no intrinsic form, unlike fruiting bodies; their structure is due to the interaction of the forces of growth of the cells and mechanical resistance of the environment. The swarm has little co-ordination except in myxobacteria and Proteus.

REFERENCES

1. KRZEMIENIEWSKI, H. & S. (1926). *Acta Soc. Bot. Pol.* **4**, 1.
2. KRZEMIENIEWSKI, H. & S. (1928). *Acta Soc. Bot. Pol.* **5**, 46.
3. LEV, M. (1954). *Nature, Lond.* **173**, 501.
4. STANIER, R. Y. (1942). *J. Bact.* **44**, 405.
5. KLIENEBERGER-NOBEL, E. (1947). *J. Hyg., Camb.* **45**, 410.
6. KVITTINGEN, J. (1949). *Acta Pathol. Belgr.* **26**, 24 & 855.
7. LOMINSKI, I. & LENDRUM, A. C. (1947). *J. Path. Bact.* **59**, 355.
8. PREUSSER, H. J. (1958). *Arch. Mikrobiol.* **29**, 17.
9. FISCHER, A. (1897). *Vorlesungen über Bakterien.* Liepzig.
10. HENRICI, A. T. & JOHNSON, D. E. (1935). *J. Bact.* **30**, 61.
11. PRINGSHEIM, E. G. (1949). *Biol. Rev.* **24**, 200.
12. BISSET, K. A. (1938). *J. Path. Bact.* **47**, 223.
13. BISSET, K. A. (1939). *J. Path. Bact.* **48**, 427.
14. BISSET, K. A. (1939). *J. Path. Bact.* **49**, 491.
15. FORSYTH, W. G. C. & WEBLEY, D. M. (1948). *Proc. Soc. appl. Bact.* **34**.
16. KLIENEBERGER-NOBEL, E. (1947). *J. gen. Microbiol.* **1**, 22.
17. BALDACCI, E., GILDARDI, E. & AMICI, A. M. (1956). *G. Microbiol.* **6**, 512.
18. BISSET, K. A. (1959). *Prog. ind. Microbiol.* **1**, 31.
19. SAITO, H. & YONOSUKE, I. (1959). *Cytologia*, **23**, 496.

CHAPTER 9

The Evolutionary Relationships of Bacteria

MORPHOLOGICAL EVIDENCE IN SYSTEMATICS

Whereas in other biological sciences the study of morphology and of systematics, has for the most part, been conducted by the same people, this is by no means true of bacteriology. Although acknowledging the importance of morphological studies, the systematists have, with a few, notable exceptions, rarely been engaged in them. Many have indeed failed to acquire sufficient familiarity with the subject to enable them fully to understand or evaluate cytological information where it is available. On the other hand, many morphologists have been so closely engaged in the study of a single character, in a very small range of morphological types of bacteria, that they have been equally badly placed to attempt the formulation of a general scheme.

There is no doubt, however, that one of the most valuable contributions of the study of bacterial cytology to biological science is the information that it affords upon numerous vexed problems of systematics. By its use an evolutionary system of classification, comparable with, and relatable to that employed in all other groups, may be applied to bacteria. [1] It has often been suggested that the lack of palaeontological evidence makes a phylogenetic classification of bacteria impossible. [2, 3] This argument expresses an implicit but totally erroneous belief that evolutionary theory in other groups of living organisms is based on their fossil remains. In fact, all the main groups are classified primarily on the evidence of living forms, and there is no reason why the same methods should not be applied to the study of bacteria. Cytology must supply the evidence, where it can.

PREVIOUS SCHEMES OF CLASSIFICATION

The best known classification of bacteria in current use is that of Bergey's Manual.[3] The major defect of this system is that it is manifestly designed as a key for identification purposes, rather than as a classification proper, and although the phylogenetic principle is acknowledged it is very little employed. Bacteria are accorded the status of a Class of Plants, the Schizomycetes (Fission Fungi) and are defined in so wide a manner that only two positive characters are permitted, small size and unicellularity. Readers of this book will be aware that few bacteria are truly unicellular; and it will appear in the present discussion that these are probably not characteristic, but are morphologically degenerate. The Class Schizomycetes is grouped with two others, the viruses and the blue-green algae. This is also unsatisfactory, since viruses cannot be regarded as a natural group at all, although some of them may be of bacterial origin;[4] and since the blue-green algae seem to be unique in never having pos-

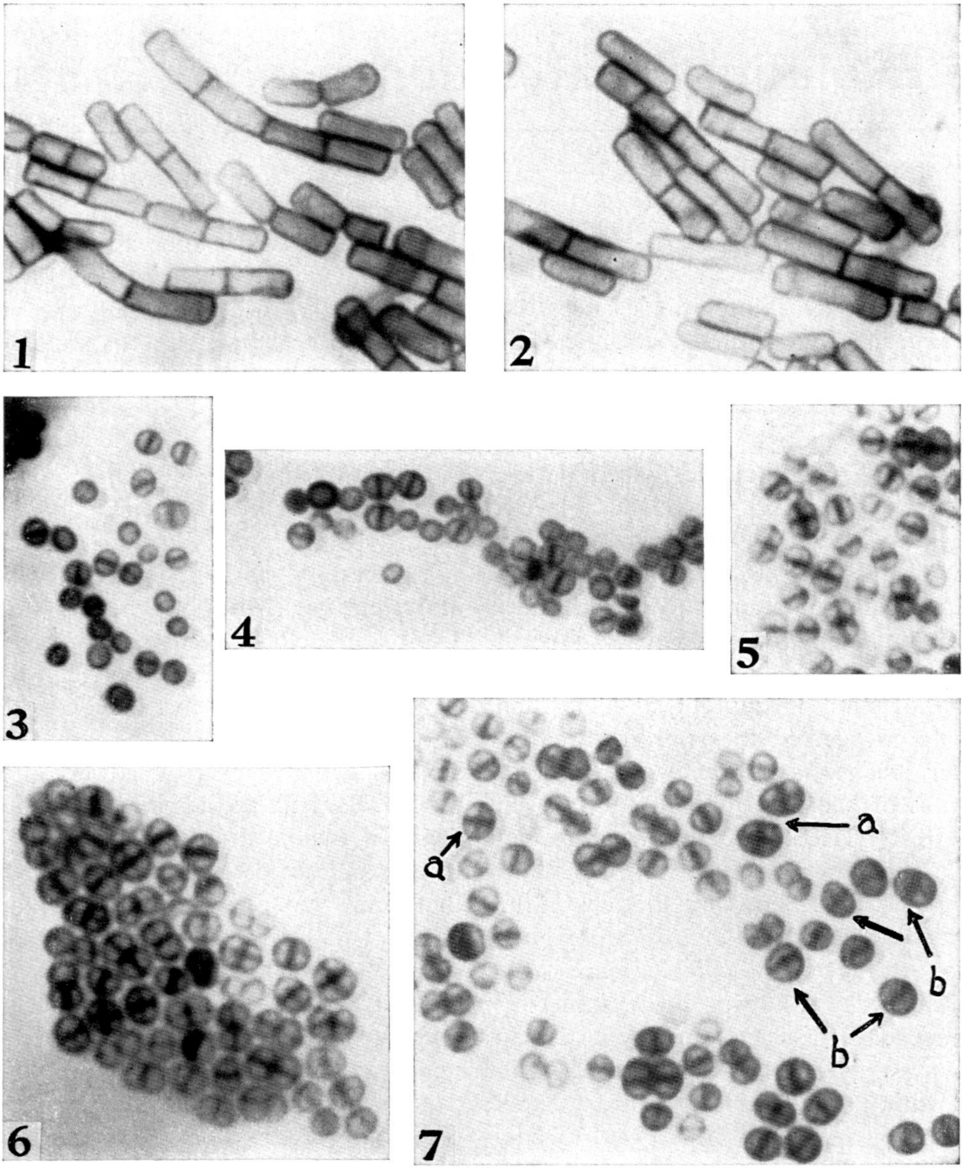
1
2
3
4
5
6
7
a
a
b
b
b

sessed flagella, which are found in plants, animals and bacteria, there is at least a case for suggesting that this relationship is too dubious to be emphasised. A circular argument develops from the inclusion with the bacteria of such obvious blue-green algae as Beggiatoa. This is a relic of the now discredited classification of chemosynthetic bacteria by purely physiological criteria, that has resulted in much confusion. [5]

There is, in fact, little reason for classifying bacteria as plants. Recent chemical analyses [6] have made it clear that their rigid cell wall (the possession of which was the main reason for the original definition) is quite unlike the cellulose wall of plants, nor does it resemble the variant types occurring in certain fungi (Chap. 3).

Two well-known schemes of bacterial evolutionary classification have been proposed in the past: that of Kluyver and van Niel [7], which is based on morphology, and that of Lwoff [8] and Knight, [9] which embodies the concept that progress of evolution is reflected in loss of synthetic ability. The first has often been quoted in general text-books, but it suffers from several logical defects, in that it proposes an independent development of spirilla, actinomycetes and sporing bacilli, from the same primitive, coccal form. Apart from the fact that this requires multiple separate origins for such basic organelles as flagella and spores, it is invalidated because the presumption of primitiveness in coccal morphology is fallacious. Cocci are, in fact, far from simple in morphology (Fig. 93), and bear a strong resemblance to sporing bacilli, from which they are probably derived, as will be explained. In contrast, the physiological schemes of Lwoff and Knight tend to bear out the hypotheses that will be derived in this chapter, from cytological premises.

PROGRESSIVE AND RETROGRESSIVE EVOLUTION IN BACTERIA

Since bacteria have flagella, and since they are structurally complex and highly adapted to various specialised modes of life, they are probably derived from the same origins as other flagellate protista, and there can be no justification for regarding them as primitive. If these conclusions are provisionally accepted, then it remains to discover how bacteria are related to the protista and to one another.

The name Schizomycetes implies that they are fungi, and it is true that Streptomyces converge morphologically with fungi, but they have the cell wall composition of bacteria. [6] Apart from this, the suggestion that the simpler bacteria may be derived by frag-

---

FIG. 93
RELATIONSHIP OF COCCI AND BACILLI

Preparations by Hale's method of the cell walls of *Bacillus*, compared with those of Gram-positive cocci. The former (1, 2) are symmetrically subdivided by cross-walls. The smaller cocci (3, 4) are normally divided by a single cross-wall; the larger cocci (5, 6, 7) are also symmetrically subdivided into four or more cells, but each cross-wall is produced at right-angles to the preceding. *a*, complete subdivision; *b*, an earlier stage of subdivision.

mentation from streptomyces-like forms of fungal origin is still unacceptable, since it once more suggests that flagella, lost for countless generations, might be regained. At the same time, while not entirely impossible, it is *a priori* improbable that an entire biological group, such as the bacteria, should have evolved from a complex terrestrial, to a simple aquatic form, whereas plants, animals and fungi alike seem to have done precisely the reverse. Aquatic Actinomycetales with motile flagellated spores have been described, [10] but their true systematological position is unproved and rather dubious.

If, on the other hand, it is considered more probable that the terrestrial bacteria are derived from the aquatic, it is obviously among the latter that the ancestral type must be sought. The most completely aquatic of all bacteria are the spirilla, and among these are to be found examples which appear to have several characters intermediate between typical bacteria and those small, saprophytic flagellates, from which it is reasonable to believe that they may have been derived.

The cell wall of spirilla is less rigid, in some cases, than is that of other bacteria, but more so than that of the flagellates. And their polar flagella, although consisting of numerous separate fibrils, almost indistinguishable from typical, monofibrillar, bacterial flagella, beat as a single organ, and take their origin, not individually, but in tufts from single blepharoplasts (Figs 32 & 34).

In other respects the spirilla are admirably suited to the role of ancestral bacteria, since they may be septate or unicellular, and are capable of forming resting cells of various types, as well as tiny, motile gonidia. But the most striking piece of evidence suggestive of a spirillar origin for bacteria is the observation of Pijper [11] that many short rod-like bacilli are slightly spiral in morphology. The conclusions that Pijper drew from this observation, especially in respect of bacterial motility, have not been accepted by other workers in this field, but the tendency to a spiral form has proved to be general in almost all bacteria except cocci.

Thus, between the spirilla and the more specialised bacteria of every type there exists a series of morphological types, through which a line of descent may be traced. The main line appears to lead to adaptation for a terrestrial environment, but genera such as caulobacteria and chlamydobacteria may also be brought into the scheme of reference, if it is concluded that they represent side-branches that have been highly evolved for a sessile, aquatic mode of life, outside the main-stream of bacterial evolution.

The series of Gram-negative forms connecting Spirillum through Vibrio and Pseudomonas with Bacterium, by a gradual simplification of the elongate spiral into a short and only slightly spiral rod, and the change from polar, through lophotrichous to peritrichous flagellation, is reasonably obvious. [12, 13] But further evolutionary progress among the more specialised Gram-negative bacteria leads on the one hand to Proteus with an enormous number of peritrichous flagella, and on the other to Aerobacter and related types some of which have discarded their flagella altogether and adopted an almost coccal morphology. It is apparent that the latter have become completely adapted to a terrestrial existence. The significance of peritrichous flagellation is less obvious; although it is often assumed that the multiflagellate swarmers of Proteus are capable of exceptionally active motility in a fluid medium, when totally immersed they swim much less actively than Vibrio with its single, polar flagellum. The swarmers, with their thousands of flagella, are notable for their ability to move upon a solid surface that is no more than moist. Under these conditions Vibrio is entirely immobilised unless the film of fluid is deep enough to permit it truly to swim. This observation, taken in conjunction with the evidence of habitat, leads to the conclusion that the profuse, peritrichous flagellation of Proteus and of certain sporing bacilli is an adaptation to life and motility, not in water, but in damp soil or decomposing organic matter. The peritrichous flagella of typical Gram-negatives represent an intermediate stage of evolution in this respect.

If any weight whatsoever can be placed upon the theory of recapitulation, the fact that the flagella in germinating microcysts of peritrichously flagellate species appear first in the polar position must be regarded as of some interest. [14] However, it need not be regarded as proven, or even probable, that existing Gram-negative bacteria are directly related one to another in this manner. What is suggested is an evolutionary trend. The actual relationships are certainly more complicated.

The most highly evolved bacteria appear to be those which have not only colonised the terrestrial habitat successfully, but have also developed adaptations for the aerial distribution of their reproductive elements. Of these the best example is provided by the streptomyces, that (presumably by evolutionary convergence) resemble minute moulds with aerial conidia some of which are covered with hairs, presumably to catch the air-currents. Although less efficient in this respect than those of the streptomyces, the spores of Gram-positive bacilli are capable of being distributed in air and dust, and it is worthy of consideration whether it may be this factor, rather than their remarkable powers of resistance, that has conferred a genetical advantage upon their possessors. It has recently been discovered that some endospores have feathery appendages [16] that can only be adapted to the purposes of aerial distribution. But as some spores are devoid of such aids, although deeply sculpted, [17] which may serve the same purpose (Figs 14, 26 & 27), it is probable that they have found them to be redundant. It is readily demonstrable that the spore-nucleus is in a condition of turgor (Chap. 4), and the refractive index of the whole structure is suggestive of dehydration. [18] These pieces of evidence, taken in conjunction with the exceedingly small relative size of most spores, and the resistance to denaturation of their proteins, suggest very strongly that the spore is reduced in size and weight by reduction of water-content, as an adaptation to aerial distribution, and that its resistance to heat and antiseptics follows as a consequence, indeed as a by-product of this process. [12, 19]

Previous concepts of the nature of spores have been based upon information, much of which is incomplete, and some definitely misleading. In the latter category, the theory that sporulation takes place in response to the stimulus of an unfavourable environment is so completely exploded as to require no further comment. A less obvious misconception, arising from a failure to realise that more or less spore-like resting cells are of universal occurrence among bacteria, is the belief that spores are especially resistant to the processes of inanition, that is to say, of oxidation, which is the main cause of death in resting cells. Obviously spores are more resistant than the corresponding vegetative cells, but there is no evidence at all to suggest that they are more so than the resting cells of non-sporing bacteria; and this is the true comparison. It is a commonplace of practical bacteriology that bacteria dried *in vacuo* will survive indefinitely.

This is a crucial point in the argument. If spores were more resistant than microcysts to inanition, then their possession would confer a most decided advantage, and no other explanation of their existence would be required. But the agencies to which spores are, in fact, especially resistant are most unlikely to be encountered under natural conditions, and, if they were, this resistance would confer little or no genetic advantage since the spore can neither metabolise nor reproduce. It may survive indefinitely in a hot spring, but evolve it cannot.

Previous theories have attempted to account for the spore in terms of its importance, not to the bacillus but to the bacteriologist, and the comparisons that have been made have not been true ones.

The development of a sporing bacillus from a spirillum, and of a streptomyces from a sporing bacillus, can be regarded as progressive evolution. The cytological and chemical resemblances between Bacillus and Streptomyces are so close that widely diverse origins cannot be attributed to them [1] and together they form the basic material of terrestrial bacteria, from which most other forms are probably derived; and almost all this further evolution has been retrogressive. In all groups of Gram-positive bacteria, it is possible to discern the same pattern of development, from morphologically complex, nutritionally unexacting, aerobic saprophytes to morphologically degenerate, nutritionally exacting, anaerobic parasites (or very specialised saprophytes), that have lost their Gram-positivity. [20, 21]

In general terms, the Actinomycetales are presumably descended from Streptomyces, and the other Gram-positive cocci and bacilli from sporing bacilli. [20] The Gram-negative bacteria, on this hypothesis, have a multiple origin, from these two Gram-positive groups and from pseudomonads. [1] Thus, the supposedly typical, Gram-negative bacteria are probably both degenerate and polyphyletic. According to the mathematical computations of taxonomic characters that have recently become popular, [22] the coliform bacteria of various genera are a small group that should be classified as species of a single genus. The sporing bacilli are much more diverse, and must be unspecialised Gram-positives, because their characters, thus analysed, show them to have a great deal in common with the Gram-negatives.

EVOLUTION OF COCCI AND ACTINOMYCETALES

Two very interesting examples of a degenerative series in bacterial evolution, such as is described in the last section, are provided by the cocci and the Actinomycetales. [20] The former appears to be derived from Bacillus, the latter, from Streptomyces.

Most like sporing bacilli are the motile, sporing sarcinae. These organisms are saprophytes, but rather specialised in that they may, for example, be adapted to the metabolism of urea, as a source both of materials and of energy. They are only slightly degenerate structurally, having lost the bilateral symmetry of the ancestral bacillus. Staphylococci are Gram-positive, but have lost both spores and flagella; they are facultative parasites that can live an almost saprophytic existence, on the skin and in animal debris. Neisseria are obligate parasites, they have lost their Gram-positivity, but are still aerobic. Veillonella are Gram-negative, anaerobic and very specialised parasites. They have lost almost all resemblance to sporing bacilli, but are nevertheless capable of being derived from them through the series of forms described above [20] (Fig. 93).

Similarly, in the case of the less complex Actinomycetales, most of which are parasites of the human and mammalian mouth, [23] a short series connects Streptomyces, through the single-spored genus Micromonospora, with Actinomyces, which is morphologically like a rather simplified Micromonospora, but is parasitic and anaerobic. [21] In this group, a second series connects the aerobic, saprophytic, morphologically complex Streptomyces with the anaerobic, parasitic, nutritionally exacting, Gram-negative Fusobacterium, which resembles a small, rather occasionally septate Bacterium. The morphology of certain Leptotrichia in the mouth is remarkably like that of the broken-up forms in which Streptomyces grows in deep culture. [21, 25] They have thick and thin filaments with occasional branches and spores; they are Gram-positive and aerobic. Other Leptotrichia are weakly Gram-positive, micro-aerophilic, do not spore, and branch only occasionally. They resemble larger, more robust specimens of Fusobacterium, with which they converge morphologically.

These are obvious examples of degenerative evolution. A similar series connects Bacillus, Lactobacillus and Streptococcus. In general, it seems that the morphologically simpler bacteria may all quite justly be suspected of being degenerate, rather than truly simple and primitive. They have lost the spores and flagella that their ancestors possessed, and have become unicellular in many cases; if not in as many cases as may still be believed, by those who have not studied their cytology.

A new definition of bacteria can thus be devised: small protista, characteristically consisting of septate filaments, rods or cocci, with unifibrillar flagella, endospores, cysts or conidia and small, motile gonidia, in some cases. All these morphological complexities are liable to be lost in the process of specialisation.

RELATIONSHIPS OF AUTOTROPHIC BACTERIA

Although inappropriate to detailed discussion in this book, it is an interesting confirmation of the validity of an evolutionary scheme for the classification of bacteria, such as has been outlined in this chapter, that the conclusions derivable from the parallel concept of progressive loss of synthetic power in the course of evolution are in excellent accordance with those based upon purely morphological reasoning. [8, 9]

This is well seen if the systematic relationships of the autotrophic bacteria are considered. Leaving aside those such as Beggiatoa, which are almost certainly not true bacteria, almost all autotrophs are either spirilla, vibrios, pseudomonads or the colonial pseudomonads, some of which are classed as chlamydobacteria. They must therefore be regarded as primitive, aquatic forms, as might reasonably be expected.

SUMMARY

There exists morphological evidence suggesting that the bacteria have evolved, in parallel with other groups of living organisms, from an aquatic to a terrestrial mode of life.

The most primitive bacteria are the spirilla,

which have characters intermediate between those of typical bacteria and flagellates.

The most highly evolved bacteria are terrestrial and have special mechanisms for the aerial distribution of their resting stages. The majority of terrestrial bacteria, especially the specialised and parasitic forms, are probably degenerate, being derived from the morphologically complex bacilli and Streptomyces. The autotrophic bacteria are relatively primitive in respect of their morphology, as they appear to be in their physiology.

REFERENCES

1. Bisset. K. A. (1962). *Symp. Soc. Gen. Microbiol.* **12**, 361.
2. Murray, R. G. E. (1960). Evolution: its science and doctrine. *Symp. R. Soc. Can.* p. 123.
3. Breed, R. S., Murray, E. G. D. & Smith, N. R. (1957). *Bergey's Manual of Determinative Bacteriology*, 7th ed. London: Ballière, Tindall & Cox.
4. Pease, P. E. & Bisset, K. A. (1962). *Nature, Lond.* **196**, 357.
5. Bisset, K. A. & Grace, J. B. (1954). *Symp. Soc. gen. Microbiol.* **4**, 28.
6. Cummins, C. S. & Harris, H. (1958). *J. gen. Microbiol.* **18**, 173.
7. Kluyver, A. J. & van Niel, C. B. (1936). *Zentbl. Bakt. ParasitKde, II*, **94**, 369.
8. Lwoff, A. (1944). *L'Evolution Physiologique. Etudes des Pertes de Functions chez les Microorganismes.* Paris: Hermann.
9. Knight, B. C. J. G. (1945). *Vitam. Horm. Lpz.* **3**, 105.
10. Higgins, M. L., Lechevalier, M. P. & Lechevalier, H. A. (1967). *J. Bact.* **93**, 1446.
11. Pijper, A. (1946). *J. Path. Bact.* **58**, 325.
12. Bisset, K. A. (1950). *Nature, Lond.* **166**, 431.
13. Leifson, E. (1960). *Atlas of Bacterial Flagellation.* New York: Academic Press.
14. Bisset, K. A. & Hale, C. M. F. (1951). *J. gen. Microbiol.* **5**, 150.
15. Baldacci, E., Gilardi, E. & Amici, E. M. (1956) *G. Microbiol.* **6**, 512.
16. Pope, L., Yolton, D. P. & Rode, L. J. (1967). *J. Bact.* **94**, 1206.
17. Bradley, D. E. & Williams, D. J. (1957). *J. gen. Microbiol.* **17**, 75.
18. Ross, K. F. A. & Billing, E. (1957). *J. gen. Microbiol.* **16**, 418.
19. Murrell, W. G. & Scott, W. J. (1957). *Nature, Lond.* **179**, 481.
20. Bisset, K. A. (1959). *Nature, Lond.* **183**, B.A. 29.
21. Bisset, K. A. & Davis, G. H. G. (1960). *The Microbial Flora of the Mouth.* London: Heywood.
22. Sneath, P. H. A. & Cowan, S. T. (1958). *J. gen. Microbiol.* **19**, 551.
23. Bisset, K. A. (1958). *Int. dent. J. Lond.* **8**, 528.
24. Bisset, K. A. (1959). *Prog. ind. Microbiol.* **1**, 31.
25. Pénau, H., Hagemann, G., Velu, H. & Peyré, M. (1954). *Revue Immunol. Thér. antimicrob.* **18**, 265.

CHAPTER 10

The Cytogenetics of Bacteria

CORRELATION WITH CYTOLOGY

Because bacteria are small, they can reproduce exceedingly rapidly. They are easily grown in artificial culture, in a limited space on inexpensive media. For these reasons, they are exceptionally suitable for genetical studies, and are widely employed for the purpose. Nowadays, the genetics of bacteria cannot properly be distinguished from genetical science in general, and many of the important new advances that have been made in the study of nuclear biology, in recent years, owe a substantial debt to bacteria as experimental material. It may be said that such medically-slanted studies as the classification of *Haemophilus influenzae* by means of its nutritional requirements, or the mechanism of sensitivity to antibiotics in Gram-positive cocci, lie at the base of our newer knowledge of protein synthesis and DNA replication.

In contrast, very few aspects of bacterial cytogenetics have been properly studied. The greatest attention has been paid to the activities of a single species, *Escherichia coli*, and to a limited number of strains. In these organisms, the replication of the nuclear molecule (Chap. 4), and the vegetative sexual conjugation process (Chap. 5) have been examined in great detail. The latter is, however, only one out of several processes observed in bacteria, that resemble conjugation, and practically no attention has been paid to the problem of correlating genetical information with what is known of the nuclear cycle in bacteria. It has, for example, been pointed out on several occasions in this book that, whereas cytologists are unanimous in describing a nuclear reduction process in the maturation of bacterial spores and other resting stages, geneticists have not examined the problem at all (Chaps 4 & 5).

Nevertheless, some degree of mutual confirmation, between genetical and cytological evidence in bacteria, does exist and has done so for many years. As early as 1948, Lederberg [1] stated that each cell of *E. coli* has more than one haploid nucleus and that after conjugation the zygote undergoes immediate reduction. This is precisely the type of nuclear cycle that was proposed for bacteria of this type, on cytological grounds, at about the same date [2, 3] (Figs 39 & 65). The genetic validity of the simple, haploid chromosome, as described by the pioneer cytologists (Chap. 4), was vindicated by Witkin [4] only three years later. Witkin's classical experiment resolved all doubt that the transverse rodlet, so convincingly visualised by Robinow, was a genuine nucleus. She showed that when young cultures, in which most cells had only a single nucleoid, were irradiated, then entire variant colonies appeared on subculture. When the cells contained two or four nucleoids, then a significant proportion of new colonies had half or quarter sectors showing the variant character. Much later, the nucleoid was confirmed to be a single molecular strand of DNA, something over a millimetre in its full, extended length. Cairns [5] who made this demonstration, used an autoradiographic technique; but the molecular thread can be seen, in its folded state, by electron microscopy of ultra-sections (Figs 27 & 38), although this gives no hint that it is a single, continuous structure, as it appears when floated out of a disrupted cell, and demonstrated by autoradiography.

Lederberg [6] also demonstrated the existence of bacterial nuclei that are genetically diploid. Morphologically, these are like multiple haploid nuclei (Fig. 54) but their minute structure has not been investigated at the time of writing. The same applies to the resting nuclei, and the apparently diploid or polyploid structures that are seen in their maturation processes, especially in the prospores of bacilli and streptomyces [3] (Figs 40 & 54). These rod-shaped nuclei, like Lederberg's diploid, continue to divide, for a longer or shorter period of time, before they undergo reduction and are transformed into the resting stage. [3] It seems that the resting stage is normally haploid, and there is no conflict between genetical and cytological evidence where they are both available for the same material. Unluckily, this is still rather exceptional.

CONJUGATION

Chapter 5 gives many examples of cytological processes in bacteria that appear to be in the nature of sexual or autogamous conjugations, but of these, only star-formation in the plant pathogens, and vegetative conjugation in *Escherichia coli* have been properly confirmed genetically, and only the latter has been studied in detail. Such conjugation was, in fact, detected by genetical analysis before it was found possible to identify the process cytologically and the experiments whereby this was achieved, by Lederberg and his collaborators, [7] are classics of microbiological history. The evidence was concerned with the nutritional requirements of bacteria. One of the commonest types of mutation in Gram-negative coliform bacteria is the loss of power to synthesise an essential cellular component. In culture, these mutant strains must be provided with all such components, in a readily available form, or they are unable to grow. Such absolute requirements (*e.g.*, for amino-acids) in the nutrition of a bacterial strain provide valuable genetic markers. Mutations occur naturally, but the rate is enormously increased by the usual mutagenic agents, such as X-rays, U.V. or mustard gas.

Commencing with a strain of *Escherichia coli*, the 'wild type', that was capable of growing upon an artificial medium containing only mineral salts, with glucose as a source both of carbon and of energy, two mutants were obtained, each of which required to be supplied with two amino-acids for its growth; although each strain was independent of the two acids required by the other. Thus, neither strain would grow on the basal medium which supported the parent wild type. However, when the two mutant strains were mixed together, and permitted an hour or so in which to interact, and were then subcultured upon the basal medium, it was found that a number of colonies appeared. These colonies consisted of bacteria capable of growing on the basal medium, without any added amino-acids, and in fact, they were genetically indistinguishable from the original, parent wild type.

It thus appeared that representatives of the two defective strains had conjugated, and that among their offspring were crosses, possessing a complete set of synthetic powers, some of which had been acquired from one parent, and some from the other. At this stage, no clear cytological evidence of mating could be detected. In cultures where conjugation was known to be taking place, it was perfectly possible to observe pairs of bacteria lying together in close proximity, but similar appearances are commonplace in microscopic preparations made from any bacterial culture, and there was no proof that they had any real meaning. This problem was solved with great ingenuity by Lederberg, [8] who obtained two defective strains that showed a high capacity to mate with one another, but were quite distinct morphologically, since one was composed of long, slender bacilli, whereas the other was notably short and plump (Figs 58 & 59). Conjugating pairs were thus very readily detectable, and they occurred much more frequently than could be accounted for by pure chance. Similar studies were made in which the two strains were marked by other visible characters, such as susceptibility to a particular strain of bacteriophage. [9]

The term sex is quite fairly applied to this process, since one strain appears to be effectively male, and passes certain parts of its genetic material to the other, which is effectively female. If the female strain is prevented from growing, after conjugation takes place, no recombinants can be found in the offspring of the male strain. [10] However, the male character can be transferred, by conjugation, to females whose offspring may subsequently be male. [11] In other words, sex in bacteria, as in animals, is a genetic character.

It was rapidly established that the hereditary characters of *E. coli* are arranged in a linear manner on a chromosome-like structure [6] and that, as this chromosome is passed from donor to recipient, the process is liable to be discontinued at almost any stage. In fact it is rarely complete although it should take, in theory, as little as half an hour, and it can be interrupted, by violent agitation of the culture, at any selected moment [12] so that the progress of transfer of genes can be studied in a controlled manner, by breaking the chromosome in selected places.

The conjugation of star-forming bacteria (Fig. 57) seems to be a very similar process, both genetically and cytologically, [13, 14] although several bacteria take part simultaneously, but the mating of *E. coli* provides the only point where the cytology and genetics of bacteria have so far come into full contact. However, so much work is now being done in both fields that further overlapping cannot long be delayed.

REFERENCES

1. Lederberg, J. (1948). *Heredity, Lond.* **2**, 145.
2. Bisset, K. A. (1949). *J. Hyg., Camb.* **47**, 182.
3. Bisset, K. A. (1951). *Cold Spring Harb. Symp. quant. Biol.* **16**, 373.
4. Witkin, E. M. (1951). *Cold Spring Harb. Symp. quant. Biol.* **16**, 357.
5. Cairns, J. (1963). *J. molec. Biol.* **6**, 208.
6. Lederberg, J., Lederberg, E. M., Zinder, N. D. & Lively, E. R. (1951). *Cold Spring Harb. Symp. quant. Biol.* **16**, 413.
7. Lederberg, J. & Tatum, E. L. (1946). *Cold Spring Harb. Symp. quant. Biol.* **11**, 113.
8. Lederberg, J. (1956). *J. Bact.* **71**, 497.
9. Anderson, T. F., Wollman, E. L. & Jacob, F. (1957). *Annls Inst. Pasteur, Paris*, **93**, 450.
10. Hayes, W. (1952). *Nature, Lond.* **169**, 118.
11. Lederberg, J., Cavalli, L. L. & Lederberg, E. M. (1952). *Genetics, Princeton*, **37**, 720.
12. Wollman, E. L., Jacob, F. & Hayes, W. (1965). *Cold Spring Harb. Symp. quant. Biol.* **21**, 141.
13. Stapp, C. & Knösel, D. (1954). *Zentbl. Bakt. ParasitKde, II*, **108**, 243.
14. Heumann, W. (1960). *Arch. Mikrobiol.* **36**, 244.

Subject Index

Subject Index

Author Index

Author Index

PRINTED IN GREAT BRITAIN BY ROBERT CUNNINGHAM AND SONS LTD., ALVA, SCOTLAND